Springer Theses

Recognizing Outstanding Ph.D. Research

Aims and Scope

The series "Springer Theses" brings together a selection of the very best Ph.D. theses from around the world and across the physical sciences. Nominated and endorsed by two recognized specialists, each published volume has been selected for its scientific excellence and the high impact of its contents for the pertinent field of research. For greater accessibility to non-specialists, the published versions include an extended introduction, as well as a foreword by the student's supervisor explaining the special relevance of the work for the field. As a whole, the series will provide a valuable resource both for newcomers to the research fields described, and for other scientists seeking detailed background information on special questions. Finally, it provides an accredited documentation of the valuable contributions made by today's younger generation of scientists.

Theses are accepted into the series by invited nomination only and must fulfill all of the following criteria

- They must be written in good English.
- The topic should fall within the confines of Chemistry, Physics, Earth Sciences, Engineering and related interdisciplinary fields such as Materials, Nanoscience, Chemical Engineering, Complex Systems and Biophysics.
- The work reported in the thesis must represent a significant scientific advance.
- If the thesis includes previously published material, permission to reproduce this must be gained from the respective copyright holder.
- They must have been examined and passed during the 12 months prior to nomination.
- Each thesis should include a foreword by the supervisor outlining the significance of its content.
- The theses should have a clearly defined structure including an introduction accessible to scientists not expert in that particular field.

More information about this series at http://www.springer.com/series/8790

YinBo Zhu

Phase Behavior of Two-Dimensional Water Confined in Graphene Nanocapillaries

Doctoral Thesis accepted by
University of Science and Technology of China,
Hefei, China

Springer

Author
Dr. YinBo Zhu
Department of Modern Mechanics
University of Science and Technology
of China
Hefei, China

Supervisor
Prof. HengAn Wu
Department of Modern Mechanics
University of Science and Technology
of China
Hefei, China

ISSN 2190-5053 ISSN 2190-5061 (electronic)
Springer Theses
ISBN 978-981-15-7956-1 ISBN 978-981-15-7957-8 (eBook)
https://doi.org/10.1007/978-981-15-7957-8

This Springer imprint is published by the registered company Springer Nature Singapore Pte Ltd.
The registered company address is: 152 Beach Road, #21-01/04 Gateway East, Singapore 189721, Singapore

Supervisor's Foreword

The Ph.D. thesis by YinBo Zhu focuses on the phase behavior of two-dimensional (2D) water confined in graphene nanocapillaries, in which some important issues on the structures and phase transitions of 2D water/ice are addressed. The simulation data and insights presented in this thesis should be significant for understanding the phase behavior of low-dimensional water/ice.

"What is the structure of water" is one of the 125 most challenging scientific questions proposed by *Science* journal in 2005. In a constrained environment, water can form new solid phases not seen in the open air. In 2015, I and co-workers demonstrated that when water is constrained between two sheets of graphene, it becomes a 2D liquid and then is turned into an intriguing monolayer solid with a square pattern under high lateral pressure. From a mechanics point of view, this liquid-to-solid transformation characterizes the compression limit (or metastability limit) of the 2D monolayer water. The graphene confinement-induced van der Waals pressure is on the order of GPa magnitude. Such a ultrahigh pressure in the confinement changes the usually tetrahedral hydrogen-bonding liquid water into quasi-2D crystal ice. It has aroused great interest in explorations on the structures and phase transitions of 2D confined water.

Although low-dimensional water/ice has been widely investigated in the past two decades, a comprehensive study on the compression limit and superheating limit of 2D water/ice under ultrahigh lateral pressure is still lacking. In this field, most of previous studies focused on the changes of temperature and confinement space, while it is still unclear that how lateral pressure influence the structure and phase transitions of 2D water/ice. The metastability limit extension of the phase diagram of liquid water at high compression/tensile pressure or high degree of superheating/supercooling can provide new features that manifest themselves in the equilibrium properties of water. It is of both fundamental and technological significance to investigate the metastability of highly constrained water in 2D confinement, including the compression and superheating limits of water confined in nanocapillaries, the structural multiformity and particularity of monolayer, bilayer, and trilayer crystalline/amorphous ice polymorphs, as well as the lateral pressure and system temperature conditions for gradual and abrupt freezing and melting.

Toward this end, the thesis by YinBo Zhu performed molecular dynamics (MD) simulations to systematically investigate the phase behavior of 2D water confined in graphene nanocapillaries. In this thesis, the author and co-workers built the mechanical model of 2D confined water in graphene nanocapillaries to investigate the structures and phase transitions of 2D constrained water under different thermodynamic conditions. A myriad of 2D ice phases and associated phase transformations were observed from MD simulations. The results demonstrated that lateral pressure can be as a primary controlling parameter to realize the first-order and continuous-like phase transitions in 2D confined water/ice. The compression limit and superheating limit (phase) diagrams of 2D water/ice in graphene nanocapillaries were obtained based on extensive MD simulations at numerous thermodynamic states as well as on the Clapeyron equation. The conclusions and insights achieved in this thesis would enrich the knowledge and cognition in the field of low-dimensional confined and interfacial water/ice. This thesis is of interest to students and scientists working in micro- and nanomechanics, condensed matter physics, and chemical physics.

Hefei, China
June 2020

HengAn Wu

Abstract

Water and ice are omnipresent on Earth and ubiquitous on comets, planets, and planet moons. On Earth, unconstrained water typically exhibits vapor, liquid, and solid (ice I_h) forms, depending on the temperature. In water/ice, the strength and directionality of the hydrogen bond combine with molecular polarity in part that makes this molecule magical and particular different with almost any other, endowing water/ice its complicated and still poorly understood phase diagram. Indeed, under a different thermodynamic environment, bulk water/ice can exist in a large number of polymorphous or polyamorphous (glassy) states.

However, in a constrained environment, water can form new solid phases not seen in the open air. Less obvious but undoubtedly more ubiquitous is water confined in microscopic environments. The special feature of confined water stems not only from intrinsic hydrogen-bonding network but also from inhomogeneity of the confined liquid, particularly in the nanoscale confinement. Besides the known thermodynamic variables, *i.e.*, temperature and lateral pressure, the structural, dynamic, and thermodynamic properties of highly constrained water are also manifestation of additional variables such as the confinement geometry as well as the physical and chemical properties of the surfaces. A better understanding of the effects of various thermodynamic variables on the phase behavior of confined water will further the advancement of scientific researches and technological applications.

When constrained in nanoscale confinement and under ultrahigh pressure, water and ice manifest dramatic differences from their bulk counterparts in part because the van der Waals pressure can induce water-to-ice transformation, known as the metastability limit of the two-dimensional (2D) liquid. From mechanics point of view, this liquid-to-solid transformation characterizes the compression limit (or metastability limit) of the 2D water. The metastability limit extension of the phase diagram of liquid water at high compression/tensile pressure or high degree of superheating/supercooling provides new features that manifest themselves in the equilibrium properties of water. In many applications, evaluation of the metastability stability (tensile/compression limit) of a solid under conditions of tension or compression is often performed to provide mechanical properties that are critical for structure design and assessment. Herein, using molecular dynamics (MD) simulations, we perform a comprehensive study of the compression/superheating limit and phase transition of 2D monolayer, bilayer, and trilayer water/ice constrained in graphene nanocapillaries.

A myriad of 2D (meta)stable ice polymorphs and polyamorphous, namely flat monolayer square-like ice (fMSI), puckered monolayer square-like ice (pMSI), monolayer triangular amorphous ice (ML-T), bilayer AB staking square-like ice (BL-ABI), bilayer AB stacking rhombic amorphous ice (BL-AB_1), bilayer AB stacking square-like amorphous ice (BL-AB_2), bilayer hexagonal ice (BL-ice I), bilayer polygonal amorphous ice (BL-A), bilayer square-tube ice (BL-VHDI), bilayer AA stacking triangular ice (BL-AAI), trilayer ABA stacking square-like ice (TL-ABAI), trilayer ABA stacking rhombic amorphous ice (TL-ABA), trilayer AAA stacking triangular ice (TL-AAAI), and trilayer AAB stacking amorphous ice (TL-AAB), are formed from the liquid water at wide ranges of nanocapillary width (*h*), lateral pressure (P_{zz}), and system temperature (*T*) in MD simulations. The compression limit (phase) diagram is obtained at the nanocapillary width versus pressure (*h*-*P*) plane, based on the comprehensive MD simulations at numerous thermodynamic states as well as on the Clapeyron equation. Interestingly, the compression limit curves exhibit multiple local minima, reflecting the metastability (phase) diagram. The inherent square characteristic of monolayer square ice is demonstrated by statistical averaging methods. Four different superheating (melting) scenarios can happen in monolayer square-like ice, demonstrating a unique and extreme sensitivity to the channel width and lateral pressure in the melting transitions. The structural transformation from BL-VHDI to BL-AAI can be regarded as the buckling failure of square ice-nanotube columns. ABA-stacked trilayer ice structures can be viewed as a bilayer clathrate hydrate with water molecules in the middle layer serving as the guest molecules. For AAA-stacked trilayer ice, the diffusion of water molecules in the layer next to the graphene is faster than that in the middle layer. Moreover, our further simulation studies reveal that lateral pressure can be used as a key controlling parameter to realize first-order and continuous-like phase transitions in the compression and superheating limits of confined 2D water/ice. The temperature threshold ($\sim$275 K) between BL-VHDI and BL-AAI is observed in the compression limit of bilayer constrained water, which indicates that the first-order and continuous-like phase transitions depend on the system temperature. In addition, in the transformations among different bilayer ices, solid-to-liquid-to-solid phase transition behavior (or Oswald staging) can arise due to that the hydrogen-bonding network must be disrupted first to reconstruct a new hydrogen-bonding network in the course of solid–solid transformation.

The essential tendency of our investigations might be significant for understanding phase behavior of 2D confined water/ice. Pressure is a fundamental thermodynamic parameter dominating the properties of materials, such as producing or predicting new materials through phase transitions under ultrahigh positive or negative pressures. It is of both fundamental and technological significance to investigate the mechanical stability of highly constrained water, including the compression and superheating limits of water confined in nanocapillaries, the structural multiformity and particularity of monolayer, bilayer, and trilayer crystalline and amorphous ice polymorphs, as well as the lateral pressure and system temperature conditions for gradual and abrupt freezing and melting.

Keywords 2D confined water · Graphene nanocapillaries · Compression limit · Lateral pressure · Square ice · Superheating · Phase transition · Phase diagram · Molecular dynamics simulation structural multiplicity and particularity of monolayer, bilayer, and trilayer crystalline and amorphous ice polymorphs as well as the lateral pressure and system temperature conditions for gradual and abrupt freezing and melting.

Keywords 2D confined water · Graphene nanocapillaries · Amorphous ice · Lateral pressure · Supercooling · Superheating · Phase transition · Phase diagram · Molecular dynamics simulation

List of Publications

Journal publications during the Ph.D. period (before July 2017)
(#Contributing equally; *Corresponding author; in chronological order):

1. Y.B. Zhu; F.C. Wang; J. Bai; X.C. Zeng* & H.A. Wu*; Compression Limit of Two-Dimensional Water Constrained in Graphene Nanocapillaries. ACS Nano, 2015. 9(12): 12197–12204.
2. Y.L. Zhao#; J. Yao#; L. Xu; M.N. Mankin; Y.B. Zhu; H.A. Wu; L.Q. Mai; Q.J. Zhang & C.M. Lieber*; Shape-controlled deterministic assembly of nanowires. Nano Letters, 2016. 16(4): 2644–2650.
3. Y.B. Zhu; F.C. Wang*; J. Bai; X.C. Zeng* & H.A. Wu; Formation of Trilayer Ices in Graphene Nanocapillaries under High Lateral Pressure. Journal of Physical Chemistry C, 2016. 120(15): 8109–8115.
4. Y.B. Zhu; F.C. Wang* & H.A. Wu; Buckling failure of square ice-nanotube arrays constrained in graphene nanocapillaries. Journal of Chemical Physics, 2016. 145(5): 054704.
5. Y.B. Zhu; F.C. Wang; J. Bai; X.C. Zeng* & H.A. Wu*; AB-stacked square-like bilayer ice in graphene nanocapillaries. Physical Chemistry Chemical Physics, 2016. 18(32): 22039–22046.
6. H.L. Gao#; Y.B. Zhu#; L.B. Mao; F.C. Wang; X.S. Luo; Y.Y. Liu; Y. Lu; Z. Pan; J. Ge; W. Shen; Y.R. Zheng; L. Xu; L.J. Wang; W.H. Xu; H.A. Wu* & S.H. Yu*; Super-elastic and fatigue resistant carbon material with lamellar multi-arch microstructure. Nature Communications, 2016. 7: 12920.
7. Z.Z. He; F.C. Wang; Y.B. Zhu; H.A. Wu* & H.S. Park*; Mechanical properties of copper octet-truss nanolattices. Journal of the Mechanics and Physics of Solids, 2017. 101: 133–149.
8. Y.B. Zhu; F.C. Wang* & H.A. Wu; Superheating of monolayer ice in graphene nanocapillaries. Journal of Chemical Physics, 2017. 146(13): 134703.
9. Y.C. Wang; Y.B. Zhu*; F.C. Wang; X.Y. Liu & H.A. Wu*; Super-elasticity and deformation mechanism of three-dimensional pillared graphene network structures. Carbon, 2017. 118: 588–596.
10. H. Yu; J. Chen; Y.B. Zhu*; F.C. Wang & H.A. Wu*; Multiscale transport mechanism of shale gas in micro/nano-pores. International Journal of Heat and Mass Transfer, 2017. 111: 1172–1180.
11. M.W. Chen; H.F. Zhan*; Y.B. Zhu; H.A. Wu* & Y.T. Gu; Mechanical Properties of Penta-Graphene Nanotubes. Journal of Physical Chemistry C, 2017. 121(17): 9642–9647.
12. J. Ge#; L.A. Shi#; Y.C. Wang#; H.Y. Zhao; H.B. Yao; Y.B. Zhu; Y. Zhang; H.W. Zhu; H.A. Wu & S.H. Yu*; Joule-heated graphene-wrapped sponge enables fast clean-up of viscous crude-oil spill. Nature Nanotechnology, 2017. 12(5): 434–440.
13. J. Xia; Y.B. Zhu*; F.C. Wang & H.A. Wu*; Effect of grain boundaries on mechanical transverse wave propagations in graphene. Journal of Applied Physics, 2017. 121(21): 215105.

14. Y.B. Zhu#; F.C. Wang# & H.A. Wu*; Structural and dynamic characteristics in monolayer square ice. Journal of Chemical Physics, 2017. 147(4): 044706.
15. Y. Hou; Y.B. Zhu; X.Y. Liu; Z.H. Dai; L.Q. Liu; H.A. Wu* & Z. Zhang*; Elastic–plastic properties of graphene engineered by oxygen functional groups. Journal of Physics D: Applied Physics, 2017. 50(38): 385305.

More publications can be found in my ORCID (0000-0001-9204-9300) or homepage: http://staff.ustc.edu.cn/~zhuyinbo/

Acknowledgements

This thesis was finished under the guidance of my advisor, Prof. HengAn Wu, at the Department of Modern Mechanics, University of Science and Technology of China (USTC). During the four years (from 2013/09 to 2017/06) at USTC, Prof. Wu gave me very meticulous guidance in both scientific research and daily life. Prof. Wu is modest and gentle, and his breadth of knowledge fill me with admiration. Prof. Wu has made a lot of achievements in the field of micro- and nanomechanics. His broad perspective, rigorous scholarship, and keen insight influence my early academic career greatly. Prof. Wu has created a good learning and research atmosphere for the laboratory. It is my great honor to join Prof. Wu's group.

Prof. FengChao Wang at USTC and Prof. Xiao Cheng Zeng at University of Nebraska-Lincoln also provided considerable help and guidance during my early research. They are co-supervisors of my thesis and several papers on confined water. Prof. Wang taught me molecular dynamics simulations and the knowledge of confined water. Prof. Zeng is an authoritative expert in the field of low-dimensional water and ice. I am thankful for their discussion and guidance.

Prof. XiuXi Wang at USTC also offered valuable advice and helped me proof-reading this thesis. He is a respectable elder and good man. Prof. Ping Gu, Dr. HuaiLing Gao, Dr. LiBo Mao, Dr. Jin Ge at USTC helped me on the experimental study of micro- and nanostructural materials. I would like to thank Dr. Weiduo Zhu at USTC for the discussion and cooperation with her. Prof. LingHui He and Prof. Yong Ni at USTC helped me during the WCCM XII and APCOM VI conference at Seoul. Prof. ZhongRong Niu at Hefei University of Technology helped me during the SCCM conference at Kunming. I am thankful for their help and discussion.

I would like to thank the members in our laboratory for their help and care. Here, I listed their names as follows: Dr. Han Wang, Dr. Fang Shi, Dr. FengPeng Zhao, Dr. YongKuan Shen, Dr. XiaoYi Liu, Dr. Jie Chen, Dr. XiaoLong Wang, Dr. LiYa Wang, Dr. YingQi Li, TianWu Zhao, Dr. Chuang Liu, Dr. Yang Cai, Dr. XiaoHao Sun, Dr. YanZi Yu, Rui Dai, Jun Xia, FengChao Wu, JinCun Fan, XiangZhe Li, Yuan Hou, Xiao Xu, YongChao Wang, ZeZhou He, Hao Yu, LiChuan Zhou,

YuanTong Li, XuLiang Qian, JiaNan Wang, MingWei Chen, Bao Wu, JiaNing Zhang, XinHeng Yuan, FanPing Sui, ZhangHui Wu, LinFeng Bu, and Ning Zhou. I would like to also thank my classmates for their help and care. Here I listed their names as follows: YongGui Sun, ZhenXian Luo, YanFei Su, HuiDong Wang, Ning Zhuang, Jian Zeng, ZhengYu Cai, ShiWei Du, Wei Tian, SiMing Chen, Tao Xi, MingHai Yu, ZhiFeng Xie, LongHui Ding, KaiTao Hua, LiuYing Shen, YaFei Zhu, ZhengHao Jin, and YuanHang Shi. I would like to also thank my old friends for their help and care. Here I listed their names as follows: XiYang Liu, Ran Miao, Kai Zhang, YuJie Li, Wen Wang, RanRan Chen, XueSong Ren, Jun Xu, Nan Zhang, and Song Luo.

At last, I would like to express my heartfelt thanks to my parents for their continued support and great love. My parents are ordinary, but they are always my foundation and encourage me to face life and move forward bravely.

The investigations presented in this thesis were jointly supported by the Strategic Priority Research Program of the Chinese Academy of Sciences (XDB22040402), the National Natural Science Foundation of China (11525211 and 11802302), and the Anhui Provincial Natural Science Foundation (1808085QA07).

May 2020 YinBo Zhu

West campus, University of Science
and Technology of China
Hefei, China

Contents

Chapter 1
Introduction

1.1 Water and Its Concept

Water is closely related to the development of human civilization [1]. The birthplace of ancient mankind civilization is in the river basin with appropriate water resources [2, 3]. For example, ancient Egyptian civilization originated from the Nile Valley. In ancient civilization, river basin and its irrigation system greatly promoted the development of Agriculture [4, 5]. Therefore, human beings could settle down and gradually prosper and develop, and then human culture emerged based on those external favorable factors. Water is a very important foundation and medium in the process of human understanding the world. To some extent, it would germinate and even promote the development of philosophy. The culture of many ancient civilizations regards water as the material basis of all things in nature world. Ancient Greek philosophers believed that water was the origin or primordial basis of all things. In ancient Chinese culture, water is regarded as one of the five elements of the origin of everything. Thus, water is closely related to the survival and development of human beings and permeates with philosophical cognition and culture.

In natural world, water is a kind of transparent, almost colorless and tasteless substance. It is the main component of rivers, lakes and oceans on Earth. It is also an indispensable and important component of organisms, known as the source of life. The chemical formula of water is H_2O, containing two hydrogen atoms and one oxygen atom with the connection of O–H covalent bond. Water molecule is a typical polar molecule. Strictly speaking, water is a liquid substance at standard ambient temperature and pressure, while the generalized concept of water also contains its solid state (ice) and gas state (steam). The existing forms of water are very common in Nature [6–8], such as rivers, lakes, oceans, frost, snow, glaciers, floating ice, clouds, rain, dew, underground aquifers, combined water, biological fluid, and various forms of interface and confined water, etc. From a biological point of view, water is essential

Y. Zhu, *Phase Behavior of Two-Dimensional Water Confined in Graphene Nanocapillaries*, Springer Theses,
https://doi.org/10.1007/978-981-15-7957-8_1

for life [9]. In organisms, water is an important solvent. Many substances in the biological fluid need water environment, and water is an important participant and medium in various metabolic processes [9–11].

An abundance of water on Earth renders its involvement in a vast number of processes that have important relevance to many fields of science. In spite of its simplicity as a liquid, different phases and anomalies still hide many unraveled secrets of water. Bulk water exhibits an unimaginably rich and complex phase diagram because of the flexibility of hydrogen bond, which is likely to maintain its mystique as the most anomalous and least understood liquid. Under different thermodynamic conditions (e.g., temperature and pressure), three-dimensional freezing water exists in a large number of polymorphous or polyamorphous (glassy) states, including at least 18 crystalline and three amorphous polymorphs [12–21].

1.2 Phase Behavior of Low-Dimensional Water

Low-dimensional confinement can be defined as the geometric space that exhibits at least one dimension with the characteristic length of several molecular diameters. Since the hydrogen-bonding network of water molecules in the low-dimensional confinements can no longer be the tetrahedral configuration, the structure of confined water is more different from that of bulk water. In addition to water clusters, the most typical low-dimensional spaces of confined water can be simplified as the quasi one-dimensional (1D) nanotube and the quasi two-dimensional (2D) nanoslit formed between two parallel walls. In the nanotubes with different diameters, confined water can form monomolecular water chains, polygonal ice nanotubes, helical ice nanotubes, multilayer ice nanotubes, etc. In the nanoslits with different channel width, confined water can form a variety of monolayer and bilayer structures. In the inhomogeneous quasi-2D confined space, confined water can form coexistence structure with different number of ice layers or solid-liquid coexistence phase.

1.2.1 Quasi-1D Confined Water

Single-wall carbon nanotube (CNT) can be seen as a thin cylindrical shell, its inner space forms the quasi-1D nanochannel. The nanoscale confinement formed in CNT can drastically change the structure and phase behavior of water. The investigations of the phase behavior and flow of confined water in CNTs are not only of great significance in basic scientific research, but also of great potential application value in material transport, desalination, nano devices, energy and chemical industry [22–30]. In 2001, using molecular dynamics (MD) simulations, Hummer et al. [31] revealed that water molecules can quickly pass through thin CNTs. As shown in Fig. 1.1, water molecules inside the CNT can form a monomolecular water chain, which is the main

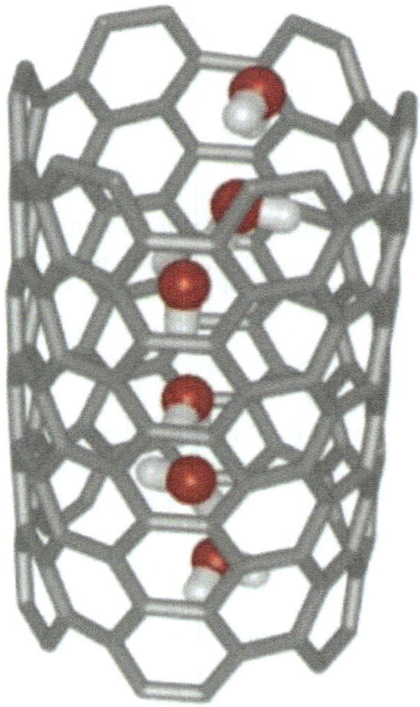

Fig. 1.1 Water chain formed in the CNT. The origin data is exported from Ref. [31]

cause of the observed fast transport. Subsequently, Koga et al. [32] found that quasi-1D confined water can form square, helical pentagonal, and hexagonal ice nanotubes inside CNTs with the diameters of 11.1 Å, 11.9 Å, and 12.6 Å, respectively.

The three ice nanotubes shown in Fig. 1.2 exhibit perfect hydrogen-bonding networks, which satisfy the ice rule that each water molecule forms four hydrogen bonds with its four neighbor water molecules. Water molecules in square and hexagonal ice nanotubes show AA-stacked molecular arrangement along the CNT, while water molecules in helical pentagonal ice nanotube show the regular helical stacking

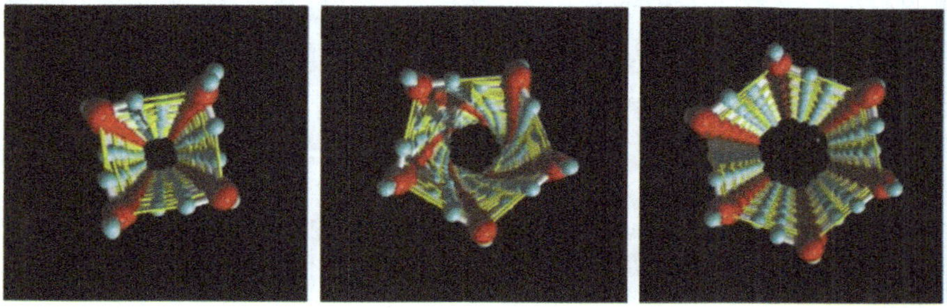

Fig. 1.2 Different ice nanotubes inside CNTs. The origin data is exported from Ref. [32]

[32]. During the phase transformation between ice nanotubes and liquid phase, potential energy curves always display obvious sudden changes, which should be mainly attributed to the formation and damage of hydrogen-bonding network in the cooling and heating processes. In some scenarios, obvious hysteresis loop can be found in the potential energy curves of cooling and heating processes, indicating the typical first-order phase transition between ice nanotubes and their liquid states [32].

Koga and co-workers continued to study the formation of ice nanotubes in single-wall CNTs. They found nine kinds of stable ice nanotubes inside CNTs with the diameters ranging from 9.0 to 17.0 Å [33]. Koga and his co-workers also gave the functional relationship between the melting point and the CNT diameter of those ice nanotubes [33]. For each type of ice nanotubes, the melting point curve has a local maximum. In their investigations, water molecules can spontaneously form ladder-like water chains in the CNT with the smallest diameter. With the increasing of CNT diameter, water molecules can form polygonal ice nanotubes and multilayer ice nanotubes. In regular polygonal ice nanotubes, the configuration of water molecules can be seen as a roll-up square net [33]. In the heating and cooling processes, all of those ice nanotubes display sudden change and hysteresis in potential energy curves, indicating the typical first-order phase transition [32, 33].

Zeng and his co-workers investigated the multilayer ice nanotubes inside large-diameter CNTs under high pressure [34]. As shown in Fig. 1.3, they found six different high-density ice nanotubes. In the CNT with the diameter of (17, 0), water molecules can spontaneously form a quasi-1D high-density ice nanotube with the double-wall helical structure, in which the outer layer is similar with the double-helix chain structure of duplex DNA while the inner layer is a four-chain helical structure. In the CNTs with the diameters of (20, 0) and (22, 0), water molecules can form four different double-wall ice nanotubes. In the CNT with the diameter of (24, 0), confined water can spontaneously freeze into a trilayer helical ice nanotube, in which the outer layer is a eighteen-chain helix and the middle and inner layers are the six-chain helical structure [34].

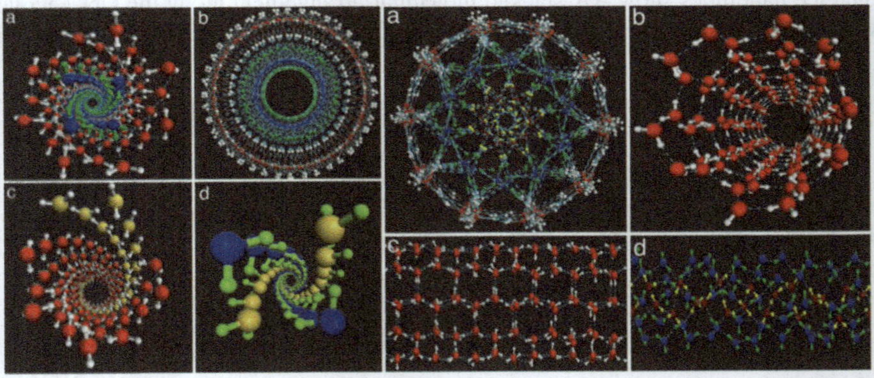

Fig. 1.3 Multilayer ice nanotubes in CNTs. The origin data is exported from Ref. [34]

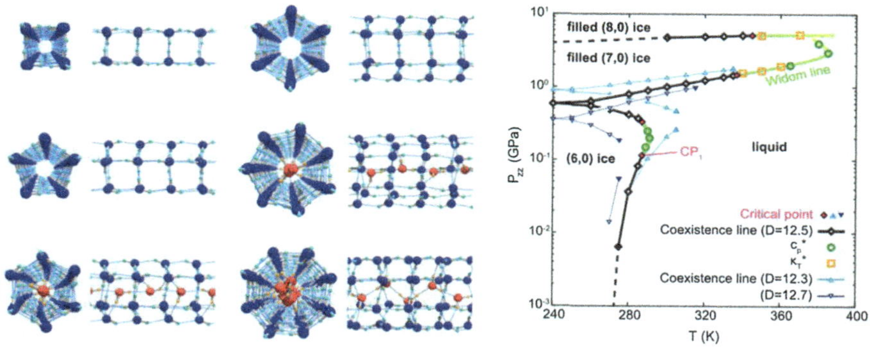

Fig. 1.4 Solid-liquid critical behavior of water confined in CNTs. The origin data is exported from Ref. [35]

Using MD simulations, Mochizuki and Koga [35] found that water confined in CNTs exhibits the solid-liquid critical behavior and multiple solid-liquid critical points. It is generally believed that the solid-liquid critical point does not exist in water, which is mainly due to the famous symmetry debate and the lack of effective experimental evidence. Nanoscale confined water can transform into low-dimensional ice phases, and the melting points of quasi-1D ices might increase obviously with the reduction of nanotube diameter. The transformations of nanoscale confined water into low-dimensional ice phases can be realized by either the first-order phase transition or the continuous-like phase transition, indicating the possible existence of solid-liquid critical point in confined water/ice systems at nanoscale. In the single-wall CNTs with some specific diameters, Mochizuki and Koga obtained the boundary of first-order solid-liquid phase transition through tracking the phase structures under different temperature [35]. They also found six different ice nanotubes as shown in Fig. 1.4a, including three different phase structures that each ice nanotube contains one inside monomolecular water chain. As illustrated in Fig. 1.4b, all of phase boundaries are terminated at the existed critical points, and they exhibit local maximum points or Widom lines. The finite-size power analysis of density distributions shows that there are both first-order and continuous-like phase transitions between these solid and liquid phases.

Recently, Zeng and his co-workers found that water confined in CNTs exists low-density and high-density liquid phases and isochore end points [36]. In supercooled bulk water, MD simulations indicate that the phase transition between low-density and high-density liquid water might exist when the temperature is lower than 230 K. While in the CNT with the diameter of 1.25 nm, the MD simulations performed by Zeng and his co-workers indicated that the low-density water and high-density water can be coexisted under ambient temperature and high pressure. In the temperature *versus* pressure phase diagram, as shown in Fig. 1.5, the regions of low-density and

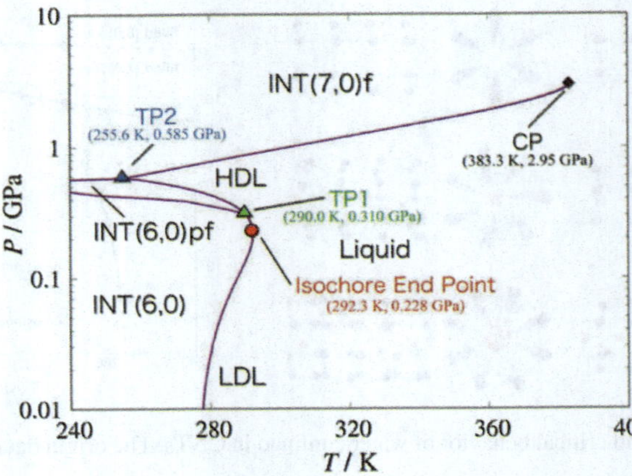

Fig. 1.5 Phase diagram of water confined in CNTs: low-density and high-density water phases and isochore end points. The origin data is exported from Ref. [36]

high-density confined water are separated by the region of hexagonal ice nanotube phase. The melting line is terminated at the isochore end point (~292 K). Under the 280 K isotherm and increasing the pressure from 10 MPa to 600 MPa, confined water inside the CNT with the diameter of 1.25 nm can occur the first-order phase transition from low-density liquid to hexagonal ice nanotube and then to high-density liquid [36].

1.2.2 Quasi-2D Confined Water

If two parallel plates are extremely closed, the gap will form quasi-2D confinement, and the characteristic length of the channel (gap) width is only a few molecular diameters. Liquid in such a highly confined space can form special quasi-2D layered structures (Fig. 1.6) [37]. The thickness of layered structures is on the order of atom-diameter magnitude. In early studies of MD simulations, hard sphere model was used to simulate the liquid molecules, and the quasi-2D confinement was formed by two parallel virtual thin plates. Schmidt and Löwen demonstrated that confinement can drastically influence the freezing phase transition and change the phase transition temperature and phase structures [38]. They also gave some quasi-2D phases, such as monolayer rhombic structure, monolayer triangular structure, bilayer triangular structure, bilayer AB-stacked structure [38]. In 1997, using TIP4P water model, Koga et al. [39] firstly revealed that confined water can transform into quasi-2D ice structure through freezing process in MD simulations. As shown in Fig. 1.7a, the bilayer ice is an AA-stacked hexagonal structure, in which the intralayer water molecules form regular hexagonal hydrogen-bonding pattern and each pair of AA-stacked water molecules are connected by the interlayer vertical hydrogen bond [39].

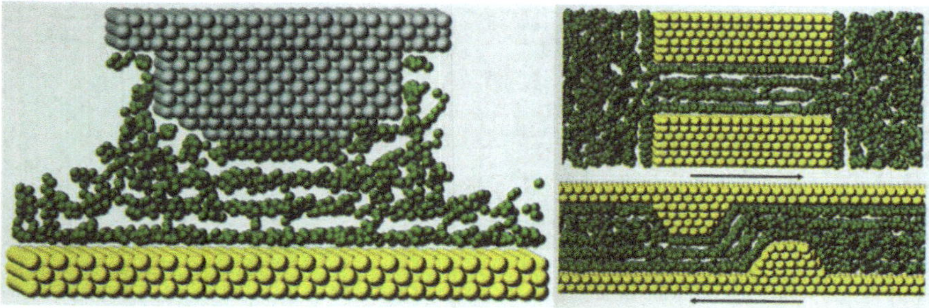

Fig. 1.6 Schematics of confined liquid. The origin data is exported from Ref. [37]

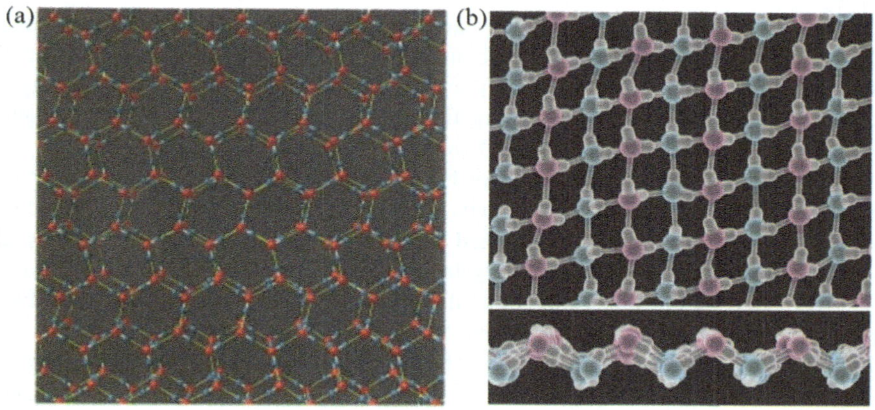

Fig. 1.7 **a** AA-stacked bilayer hexagonal ice. **b** Monolayer ice. The origin data is exported from Refs. [39, 43]

Subsequently, TIP3P, SPC, TIP4P, TIP5P, and other rigid water models were used to investigate the phase behavior and phase transition of quasi-2D confined water. In different confinement and under different thermodynamic conditions (e.g., pressure and temperature), quasi-2D confined water can transform into various 2D ice phases.

In terms of the stratification of water molecules, quasi-2D ice structures can be simply classified into monolayer ice, bilayer ice, and trilayer ice, including crystal and amorphous phases. Quasi-2D ice structures had also found in previous experimental studies without the environment of confinement. For example, in 2009, bilayer hexagonal ice was obtained at the surface of graphene/Pt substrate through the low-temperature vapor deposition method [40]. Monolayer ice could also be obtained by this method [41]. While recent experimental study also demonstrated that bilayer hexagonal ice can be formed inside the nanoslit between graphene and MoS_2 sheets [42]. The main concern of this thesis is the computer simulations on 2D water and ice in graphene nanocapillaries. Therefore, we herein do not introduce the experiments on 2D water and ice too much. Monolayer ice was firstly observed by Zangi and

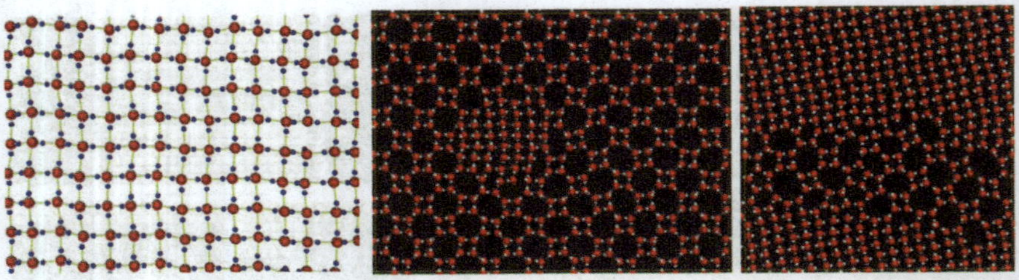

Fig. 1.8 Structures of monolayer ice and phase coexistence. The origin data is exported from Refs. [44, 45]

Mark using TIP5P water potential in MD simulations [43]. As shown in Fig. 1.7b, the monolayer ice exhibits a zigzag-like puckered rhombic structure and regular proton arrangement, which can be obtained by changing the plane separation and system temperature. The transformation between the monolayer ice the monolayer liquid water is a first-order phase transition. The coexistence of monolayer ice and liquid water was also observed in MD simulations [43].

Subsequently, some research groups have conducted in-depth research on monolayer ice and classified it according to its structure in detail. Under different temperature and pressure and in different-sized nanochannels, MD simulations manifested that monolayer ice has six different crystal phases at least [44–46]. Koga and Tanaka [44] investigated the phase diagram of monolayer water confined between two parallel hydrophobic surfaces and gave the thermodynamic properties of quasi-2D confined water. In the MD simulations of Koga and Tanaka, the lateral pressure was fixed at 0.1 MPa or 200 MPa, and the phase transitions were mainly realized through changing temperature. As illustrated in Fig. 1.8a, the monolayer ice exhibits perfect hydrogen-bonding network except some point defects [44]. The flat monolayer ice will be puckered if the channel width enlarges slightly. Zeng and his co-workers then performed in-deep investigations on monolayer ice [47, 48]. As shown in Fig. 1.8b, c, they found the phase coexistence of low-density and high-density monolayer ice structures and four different monolayer ice phases [47].

For bilayer ice, except the bilayer hexagonal ice in Fig. 1.7a, Zeng and his co-workers [49] also found the bilayer polygonal ice, bilayer square-tubes ice, and two different high-density amorphous ices in MD simulations with TIP5P force field. The bilayer polygonal ice is similar with the bilayer hexagonal ice due to the existence of pentagonal and heptagonal elements. Bilayer polygonal ice and bilayer hexagonal ice can transform into each other through the heating and cooling processes [49], and MD simulations demonstrated that bilayer hexagonal ice is a more stable structure with low energy compared with bilayer polygonal ice. Figure 1.9 shows the detailed structure of bilayer square-tubes ice [49], in which water molecules form the parallel square-tube array and adjacent ice tubes are staggered. Bilayer square-tubes ice exhibits regular hydrogen-bonding network and ordered proton arrangement, which can be transformed from bilayer water under low temperature and high lateral pressure. The transformation between bilayer square-tubes ice and bilayer hexagonal

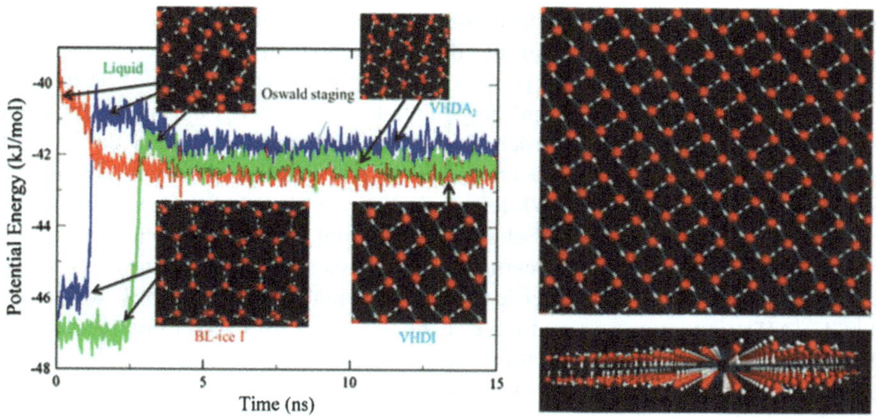

Fig. 1.9 Bilayer square-tubes ice (named as BL-VHDI) and its hydrogen-bonding network. The origin data is exported from Ref. [49]

ice can be realized through changing the lateral pressure. During the phase transition, bilayer high-density liquid phase appears as the intermediate state, known as the Oswald stage phenomenon [50]. The transformation from bilayer liquid water to bilayer square-tubes ice is a first-order phase transition. Under higher lateral pressure, Zeng and Bai also found two different high-density amorphous ice structures [49].

Stanley and his co-workers [51] performed in-deep investigations on bilayer confined water and firstly revealed the continuous-like phase transition in bilayer water. As illustrated in Fig. 1.10, low-density bilayer liquid water can transform into

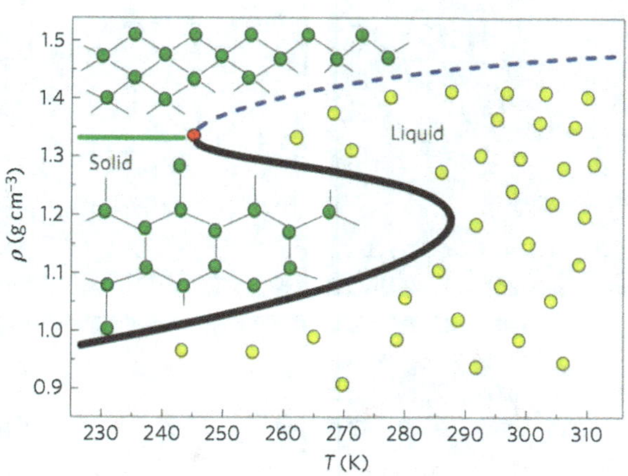

Fig. 1.10 First-order and continuous-like phase transitions in bilayer water. The origin data is exported from Ref. [51]

bilayer hexagonal ice through the first-order phase transition, while high-density bilayer liquid water can transform into bilayer rhombic amorphous ice through the continuous-like phase transition. The distinction between first-order and continuous-like phase transitions is based on the potential energy curves during heating and cooling processes. In the first-order phase transition, potential energy curves will display obvious sudden changes at critical points and appear a hysteresis loop during heating and cooling processes. While potential energy curves in the continuous-like phase transition show continuous change and without hysteresis loop [51].

Trilayer ice was rarely concerned in previous MD simulations. Relatively speaking, crystal phase in trilayer ice is few. Giovambattista et al. [52] carried out detailed MD simulations on multilayer confined water. In trilayer confined water, they found that a trilayer ice structure can be spontaneously formed when the density increases. The trilayer ice exhibits two ordered outside layers, while the middle layer is in liquid-like state. Jia et al. [53] found a trilayer ice with the hybrid structures of twisted rectangles and rhombuses (Fig. 1.11). Kumar et al. [54] found a trilayer ice that contains zigzag-like middle layer and two flat outside layers. In the same time, Kastelowitz et al. [55] found that trilayer ice can exhibit flat middle layer or puckered middle layer based on the channel width. Zangi and Mark [56] obtained trilayer ice in MD simulations through applying homogenous electric field.

Note that, all of above mentioned few-layer ice structures are the results from MD simulations before 2015. In those investigations [31–56], the phase transitions were realized mainly through changing the number of water molecules,

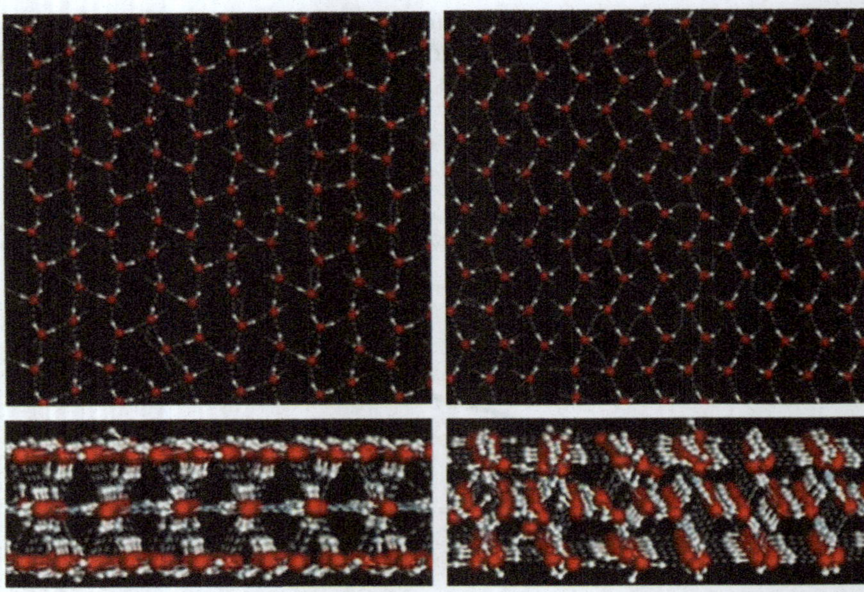

Fig. 1.11 Trilayer ice structures and hydrogen-bonding networks. The origin data is exported from Refs. [46, 53]

Fig. 1.12 Schematic of 2D water confined between two graphene sheets. The origin data is exported from Ref. [57]

size of confinement, and system temperature. Only few studies paid attentions to pressure-dependent phase transitions in confined water.

In 2015, using the high-solution transmission electron microscopy (TEM), Algara-Siller et al. [57] found that 2D square crystal structures can be formed when water droplets are trapped by two graphene sheets. In the experiments, a water nanodroplet was initially placed on one graphene sheet, and another graphene sheet was then placed on the top, covering the water droplet. The strong graphene-graphene interaction makes the two graphene sheets come together at the margin, and the van der Waals (vdW) pressure squeezes some water out of the channel while some others are trapped inside the graphene pockets and form few-layer ice. Meanwhile, the complementary MD simulations (with a helium bubble and water nanodroplet) in Ref. [57] indicated that the lateral pressure can simulate and characterize the vdW pressure in the graphene nanocapillaries (Fig. 1.12). The vdW pressure in the graphene pocket is on the order of GPa magnitude [57–59]. Such a ultrahigh pressure in the confinement changes the usually tetrahedral hydrogen-bonding liquid water into quasi-2D crystal ice. The experimental observation of square ice made it to be a hot topic and become a matter of debate. Although Zhou et al. [60] questioned that the observed square ice might be the NaCl crystal and bilayer square ice has not been found in simulations, the concept of ultrahigh vdW pressure inside graphene nanocapillary has been widely accepted and recognized by scientific community. Toward this end, in this thesis, we built the mechanical model of 2D confined water in graphene nanocapillaries to investigate the phase transitions and phase behavior of 2D constrained water under different conditions (e.g., lateral pressure, temperature, and channel width). In this Chapter, we also provided a brief literature review of recent investigations on 2D water constrained between two graphene sheets reported by other research groups, including MD simulations based on classical force fields and first-principle calculations based on density-functional theory (DFT).

Corsetti et al. and Chen et al. firstly carried out DFT calculations on monolayer ice confined between two graphene sheets [61–63]. Corsetti et al. [61] found that square network and hexagonal network in monolayer ice exhibit multiple topological structures, which were classified by the proton arrangement and molecular dipole moment. Chen et al. [62] used the DFT-based ab initio MD simulations to search different configurations of monolayer ice. As shown in Fig. 1.13a, they found five different monolayer ice structures. The regular hexagonal monolayer ice in Fig. 1.13a has been found in previous MD simulations, while the pentagonal monolayer ice and buckled square monolayer ice are firstly reported in computational simulations [62]. In Ref. [63], Chen et al. defined the flat monolayer ice combined with square and

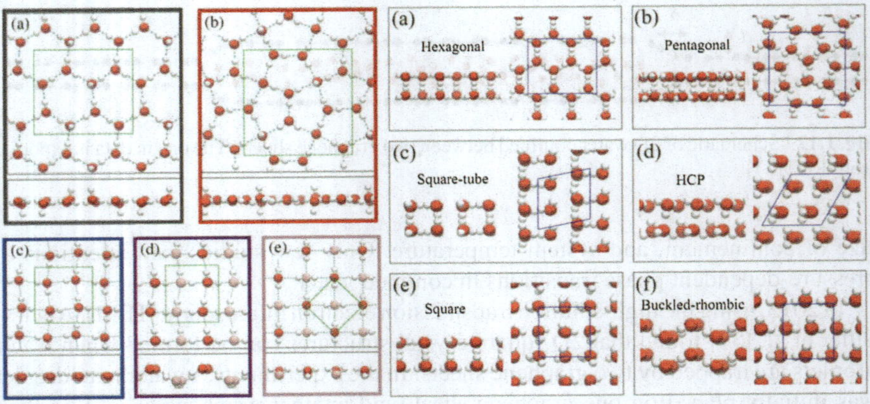

Fig. 1.13 Monolayer ice and bilayer ice obtained from first-principle calculations. The origin data is exported from Refs. [62, 68]

rhombic elements as the square phase of monolayer ice. The zigzag-like monolayer ice was defined as the rhombic phase, in which the configuration of water molecules is similar with the polarized water chains reported by Strauss et al. [64]. from MD simulations. In addition. Roman and Groß [65] found another rhombic phase of monolayer ice with the same proton arrangement for each water molecule, which is similar with the structure reported by Zangi and Mark [43]. Roman and Groß also found the rhombic-like monolayer triangular ice and the hexagonal-like prism structure [65].

Simultaneously, many research groups performed investigations on 2D ice in graphene confinement, but the stable bilayer square ice has not been obtained from computational simulations. Corsetti et al. [66] firstly carried out first-principle calculations on bilayer confined water. They found a square-wave-like bilayer rhombic ice, in which the top view is rhombic configuration and the side view is square-wave configuration [66]. However, other research groups obtained similar structure with the square feature in top view. The simulated results demonstrated that this bilayer square ice is unstable [65, 67, 68]. In bilayer ice structures obtained from first-principle calculations (Fig. 1.13b), only the configuration of bilayer hexagonal ice is the same with that predicted from MD simulations [68]. The AA-stacked bilayer pentagonal ice shows the worst stability [68], which is different from the interlocked pentagonal bilayer ice reported by Zeng and his co-workers [69]. Bilayer square-tubes ice in Fig. 1.13b is similar with the structure obtained from MD simulations [49], while the proton configurations display some fine differences among the results from different research groups [65–68]. The hexagonal-close-packed bilayer ice is similar to the bilayer triangular ice observed in MD simulations. The bilayer buckled-rhombic ice is similar to the AA-stacked two layers of rhombic ice [68].

Roman and Groß [65] found other bilayer ice phases from first-principle calculations. For example, the bilayer herringbone ice structure. Neek-Amal and his co-workers [67, 70–72] investigated the phase structures and commensurability effects of 2D confined water using MD simulations. In addition, Zeng and his co-workers [73] found the coexistence of different quasi-2D ice and the coexistence of liquid water and 2D ice in inhomogeneous confinement. Although 2D square ice in experiment still remains a matter of debate, it has aroused great interest in theoretical simulations on the structures and phase behavior of 2D confined water. There are many differences among simulation results obtained from different research groups and computational methods. MD simulations and first-principle calculations would be limited to the selected force field and density functions, respectively. The computational scale and simulation methods are different from the actual conditions in experiments, and there might be some unknown factors in experiments. Therefore, more in-depth investigations on the phase behavior of 2D confined water are needed. "What is the structure of water" is the 46th question of the 125 most challenging scientific frontier issues announced by *Science* journal [74]. The study on 2D confined water is not only of great significance in basic scientific research, but also of great application prospect in many fields.

1.3 Confinement Effect of Graphene Nanocapillary on Water

Graphene, the first 2D atomic crystal available to us, is a one atom thick and closely packed 2D sp^2-bonded carbon honeycomb lattice [75]. Due to a large number of its material parameters are supreme, such as mechanical strength and elasticity, very high electrical and thermal conductivity, etc., graphene has attracted much attention from the scientific research community [76–85]. Graphene oxide (GO) is the derivative of graphene [86], which could be regarded as that some oxygen-containing functional groups are introduced to the surface and edge of graphene. The commonly accepted structural model of GO is that hydroxyl and epoxy groups are randomly distributed on the surface, while carboxyl and carbonyl groups are introduced at the edge [86]. As a new type of carbon materials with excellent properties, GO exhibits high specific surface area and rich oxygen functional groups, rendering it to be promising in graphene-based materials [87–96]. In 2007, Ruoff and his co-workers [88] firstly obtained the GO paper (membrane) through vacuum filtration method. This laminate material is mainly composed of GO sheets with the nacre-like (brick-and-mortar) microstructure, in which the interlayer interactions are the hydrogen bonds formed among oxygen functional groups and water molecules and the van der Waals interactions between adjacent GO sheets. In 2012, Geim and his co-workers [97] found that GO film can allow the penetration of water (vapor) easily, while most of other gases are rejected (Fig. 1.14a). This anomalous mass transfer behavior under nanoscale confinement through GO film makes the investigations of graphene-based fast water transport and ion screening become the international frontiers and hot points [97–100].

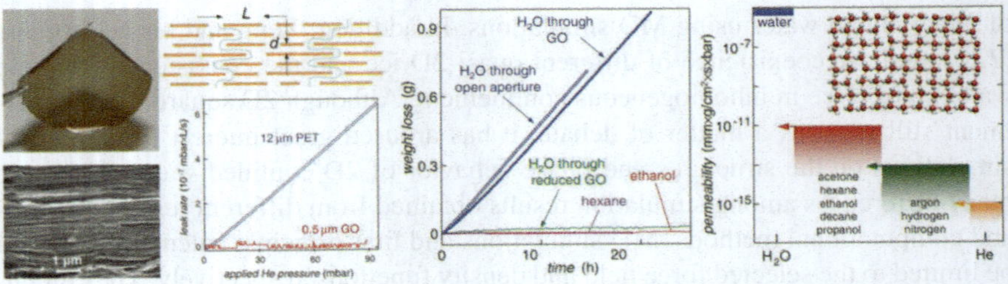

Fig. 1.14 GO film and its nanocapillaries for water transport. Fast transport of water through GO membrane and its underlying mechanism at nanoscale. The origin data is exported from Ref. [97]

MD simulations demonstrated that the combined effect of capillary pressure and evaporation results in the fast transport of water through graphene nanocapillaries (Fig. 1.14b), instead of the driving force of traditionally believed pressure difference [97]. This study revealed that the flow behavior of water inside graphene nanochannel is different from classical hydrodynamics at nanoscale, which is of significance to understand the fast transport of water in low-dimensional confinement. In GO film, the oxygen functional groups can be as the support between adjacent GO sheets, resulting in the formation of atomically smooth graphene nanocapillaries in unoxidized area between staggered sheets. The channel width of graphene nanocapillaries is the size of 2D nanoslit in laminate GO film. Moreover, the structural defects (e.g., nanopores) in GO can provide additional path for water transport. Recent theoretical study also revealed that the structure and hydrogen-bonding behavior of water molecules in oxidized and unoxidized regions as well as the channel width between graphene sheets can significantly influence the water transport [101].

Theoretical simulations indicated that water molecules can enter into graphene nanocapillary when the channel width is only 0.35 nm, in which liquid water would spontaneously transform into 2D solid-like crystal [97]. That is the monolayer square-like ice as shown in Fig. 1.15 [57]. Monolayer square-like ice is an novel form of water under ambient temperature, breaking through the long-standing understanding of ice, which demonstrates the magical configuration of water constrained in nanoscale confinement. MD simulations indicated that the vdW interaction between two graphene sheets can applied GPa magnitude capillary-like pressure to confined water [57]. This highly confined effect forced the transformation of hydrogen-bonding network from 3D structure into 2D pattern, then the regular arrangement of water molecules in 2D pattern formed the monolayer square ice. The simulations also demonstrated that this square ice in 2D nanoscale confinement could be universal and be not limited to graphene nanocapillaries; that is, monolayer square-like ice could be formed in nanochannels with different surface properties [57]. After the experimental report on monolayer square ice in Algara-Siller et al. [57], 2D square ice has become a hot topic and remains a matter of debate, especially on the structure. Subsequently, MD simulations and DFT calculations indicated that confined water in graphene nanocapillaries can form various 2D ice structures [61–73]. Although consensus has been reached on monolayer square-like ice, the results on bilayer

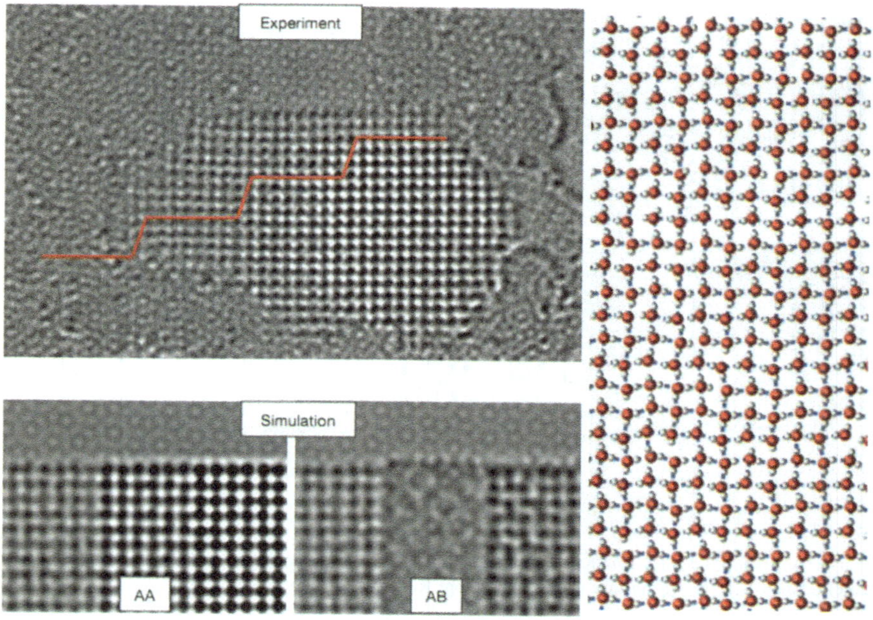

Fig. 1.15 Water molecules inside two sheets of graphene form monolayer square ice. The origin data is exported from Ref. [57]

ice still exist divergence and further theoretical and experimental investigations are needed.

In recent experimental study, Geim and his co-workers [102] fabricated graphene nanochannel with atomistic precision and atomically smooth wall. The size of graphene nanocapillaries can be controlled by the layer number of graphene spacer. In this study, researchers proposed a new method to fabricate nanocapillaries, which could stack different-sized 2D crystal materials into nanochannels just like building blocks. The size precision of this kind of nanochannels can be controlled at 0.34 nm, which is the smallest nanochannel fabricated in laboratory. The atomically smooth wall of graphene nanocapillary endows it with ultrahigh flux of water transport (~1 m/s). MD simulations revealed that disjoining pressure at molecular level significantly enhances the driving force of water transport in graphene nanocapillaries [102], rendering it vastly different size effect at nanoscale compared to macroscopic scale. The interfacial slipping between graphene and water also contributes significantly to water transport. This investigation also demonstrated that the solid-liquid interfacial interaction shows a decisive effect on the flow behavior at nanoscale (Fig. 1.16).

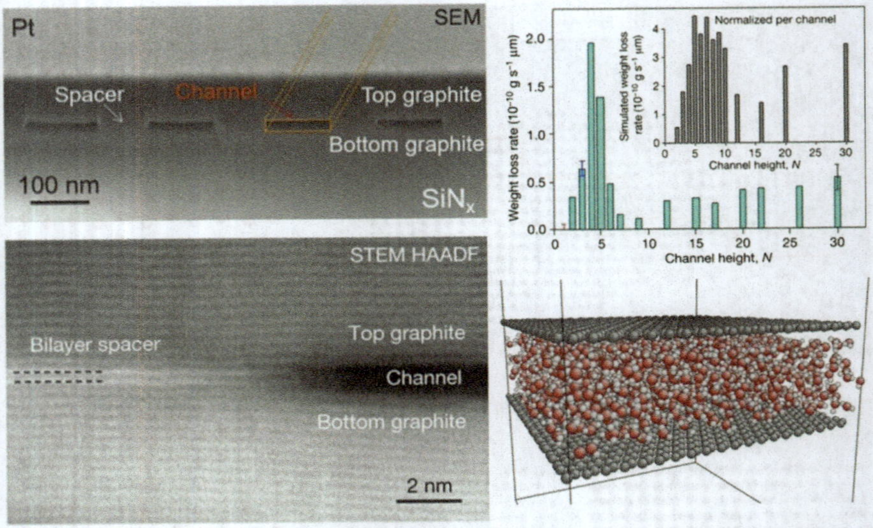

Fig. 1.16 Graphene-based nanochannel with the atom-scale precision and its size effect of fast water transport. The origin data is exported from Ref. [102]

1.4 Some Important Issues Concerned in This Thesis

1.4.1 Confinement Effect in Graphene Nanocapillaries

Recent experiments demonstrated that graphene-based nanochannel exhibits excellent water transport ability and ion selectivity [97, 103, 104]. Graphene nanochannels with atom-scale precision and smooth walls influence significantly on the understanding and cognition of mass transfer at nanoscale [102]. The size of fabricated nanochannels in experiments is comparable to the acceptable model size in computational simulations, endowing theoretical studies based on MD simulations with more practical significance. Disjoining pressure [105–107] introduced by solid-liquid interface at nanoscale provides additional driving force for mass transfer in nanochannels. The nature of disjoining pressure in nanocapillaries is the comprehensive embodiment of various intermolecular interaction at solid-liquid interface, which is pivotal to the abnormal mass transfer behavior in nanometer-sized channels. In the investigations of water in nanoscale confinement, the confinement effect-induced pressure and local ordered structure (hydrogen-bonding behavior) should be considered due to the importance of solid-liquid interfacial interaction at atomistic scale [57].

1.4.2 Structures of 2D Ice

2D water can form various crystalline and amorphous solid structures under corresponding thermodynamic conditions [46]. There still remains some differences between the structures predicted from DFT calculations and MD simulations. In recent, Algara-Siller et al. [57] observed 2D square ice from TEM experiments, while the existence of bilayer square ice still remains a matter of debate due to the lacking of solid evidence in theoretical simulations [60]. It is important to explore the formation and structural features of 2D ice phases in such a nanoscale confinement, which can provide in-depth understanding on the phase behavior of highly constrained water.

Currently, there is widespread consensus on the structure of monolayer square ice. The slightly rhomboidal feature in the simulated monolayer square-like ice is mainly due to the existence of rhombic elements in the structure. However, stable bilayer square ice at ambient temperature has not been observed from both MD simulations and DFT calculations. Perhaps, there is a possibility that current simulation level cannot capture the bilayer square ice structure or that the experimental bilayer square ice is a result of epitaxial growth out of the monolayer square ice. This might be an open question to discuss that what is the structure of water/ice under 2D limit. The discrepancy between simulations and experiments requires further investigation from both theoretical modelling and experimental measurements.

We should also note that a myriad of 2D ice structures have been predicted from computational simulations and some 2D ice phases have been demonstrated by experiments [39–56]. The exploration of 2D water/ice structures should be meaningful and significant in many fields of science, such as physics, chemistry, geoscience, nanoscience, biological science, etc.

1.4.3 Phase Transitions in 2D Water/Ice

Pressure is a basic thermodynamic variable, which can greatly affect the state and properties of matter. It is thus an important method for the formation of intriguing materials not seen at ambient conditions [108]. Algara-Siller et al. [57] recently demonstrated that when water is constrained between two sheets of graphene, it becomes a 2D liquid and then is turned into an intriguing monolayer solid with a square pattern under high lateral pressure. From a mechanics point of view, this liquid-to-solid transformation characterizes the compression limit (or metastability limit) of the 2D monolayer water. Thus, lateral pressure plays an important role in the phase transition from liquid water to monolayer square ice in graphene nanocapillaries. In previous studies on confined water, first-order and continuous-like phase transitions have been demonstrated through the changing of temperature and channel width [44, 49, 51, 109]. In recent, some special phase transitions and transformation pathways were observed in both 1D and 2D confined water/ice systems, such as the solid-liquid critical behavior [35], isochore end point [36, 110], and weak first-order transition [36, 110]. 2D ice can be seen as an special solid material. Evaluation of the tensile/compression limit of a solid under conditions of tension or compression is often performed to provide mechanical properties that are critical for structure design and assessment. The investigation on phase transitions in 2D water/ice can reveal some unknown extraordinary characteristics of 2D materials and might open up a novel perspective for understanding the phase behavior in low-dimensional water/ice systems [111].

1.5 Thesis Overview

This thesis is organized as follows.

Chapter 1 Introduction

The background of this thesis is briefly introduced. A literature review on the structures and phase transitions of low-dimensional water/ice is carried out.

Chapter 2 Methods and Models

The comparison among different water models (force fields) is briefly introduced. Computational methods and models used in this thesis are presented in detail.

Chapter 3 Monolayer Square-like Ice between Two Graphene Sheets

The structural and dynamic characteristics of monolayer square-like ice in graphene pocket are investigated in detail. The inherent square pattern of monolayer square-like ice in MD simulations is demonstrated by statistical averaging methods with geometry- and time-averaging effects.

Chapter 4 Superheating Behavior of Monolayer Ice in Graphene Nanocapillaries

Phase transitions of monolayer ice during the superheating process are investigated. Four different melting scenarios and monolayer triangular amorphous ice are observed during MD simulations.

Chapter 5 AB-Stacked and AA-Stacked Bilayer Ices in Graphene Nanocapillaries

Bilayer phases of 2D water constrained in graphene nanocapillaries are discussed, including the AB-stacked and AA-stacked structures. Phase transitions among different bilayer ices were also discussed in detail.

Chapter 6 Trilayer Ice in Graphene Nanocapillaries

Three different trilayer ice phases are observed in MD simulations. The structural feature, phase transition and dynamic behavior of trilayer water in graphene nanocapillaries was systematically studied.

Chapter 7 Compression Limit of 2D Water in Graphene Nanocapillaries

The compression limit of 2D water confined in graphene nanocapillaries is systemically investigated, and the associated phase diagram is obtained.

Chapter 8 Summary and Future Work

This chapter summarizes the work in this thesis. We also give the perspective of some significant yet unresolved issues at the end of this thesis.

References

1. Priscoli JD (2000) Water and civilization: using history to reframe water policy debates and to build a new ecological realism. Water Policy 1(6):623–636
2. Solomon S (2010) Water: the epic struggle for wealth, power, and civilization. Harper Collins
3. Gibson MG (1974) Violation of fallow and engineered disaster in Mesopotamian civilization. Irrigation's impact on society. Univ Arizona Anthropol Pap 25:7–20
4. Ayers RS, Westcot DW (1985) Water quality for agriculture. Food and Agriculture Organization of the United Nations, Rome
5. Viala E (2008) Water for food, water for life a comprehensive assessment of water management in agriculture. Irrigat Drain Syst 22(1):127–129
6. Robert F (2001) The origin of water on Earth. Science 293(5532):1056–1058
7. Govender M, Chetty K, Bulcock H (2007) A review of hyperspectral remote sensing and its application in vegetation and water resource studies. Water Sa 33(2):145–151
8. Morbidelli A, Chambers J, Lunine JI et al (2000) Source regions and timescales for the delivery of water to the Earth. Meteorit Planet Sci 35(6):1309–1320
9. Franks F (2000) Water: a matrix of life. Royal Society of Chemistry
10. Luck WAP (1976) Water in biological systems: topics in current chemistry. Springer, Berlin
11. Karr JR (1991) Biological integrity: a long-neglected aspect of water resource management. Ecol Appl 1(1):66–84
12. Lobban C, Finney JL, Kuhs WF (1998) The structure of a new phase of ice. Nature 391(6664):268–270
13. Salzmann CG, Radaelli PG, Hallbrucker A et al (2006) The preparation and structures of hydrogen ordered phases of ice. Science 311(5768):1758–1761
14. Falenty A, Hansen TC, Kuhs WF (2014) Formation and properties of ice XVI obtained by emptying a type sII clathrate hydrate. Nature 516(7530):231–233
15. Del Rosso L, Celli M, Ulivi L (2016) New porous water ice metastable at atmospheric pressure obtained by emptying a hydrogen-filled ice. Nat Commun 7:13394
16. Mishima O, Calvert LD, Whalley E (1984) Melting ice I at 77 K and 10 kbar: a new method of making amorphous solids. Nature 310(5976):393–395
17. Mishima O, Calvert LD, Whalley E (1985) An apparently first-order transition between two amorphous phases of ice induced by pressure. Nature 314(6006):76–78
18. Hemley RJ, Chen LC, Mao HK (1989) New transformations between crystalline and amorphous ice. Nature 338(6217):638–640
19. Mishima O, Stanley HE (1998) The relationship between liquid, supercooled and glassy water. Nature 396(6709):329–335
20. Tse JS, Klug DD, Tulk CA et al (1999) The mechanisms for pressure-induced amorphization of ice Ih. Nature 400(6745):647–649
21. Loerting T, Salzmann C, Kohl I et al (2001) A second distinct structural "state" of high-density amorphous ice at 77 K and 1 bar. Phys Chem Chem Phys 3(24):5355–5357
22. Ghosh S, Sood AK, Kumar N (2003) Carbon nanotube flow sensors. Science 299(5609):1042–1044

23. Mikami F, Matsuda K, Kataura H et al (2009) Dielectric properties of water inside single-walled carbon nanotubes. ACS Nano 3(5):1279–1287
24. Maniwa Y, Matsuda K, Kyakuno H et al (2007) Water-filled single-wall carbon nanotubes as molecular nanovalves. Nat Mater 6(2):135–141
25. Lee CY, Choi W, Han JH et al (2010) Coherence resonance in a single-walled carbon nanotube ion channel. Science 329(5997):1320–1324
26. Reiter G, Demirel AL, Granick S (1994) From static to kinetic friction in confined liquid films. Sci-AAAS-Wkly Paper Ed-Includ Guide Sci Inf 263(5154):1741–1743
27. Rasaiah JC, Garde S, Hummer G (2008) Water in nonpolar confinement: from nanotubes to proteins and beyond. Annu Rev Phys Chem 59:713–740
28. Liu S, Luo J, Li G et al (2008) Effect of surface physicochemical properties on the lubricating properties of water film. Appl Surf Sci 254(22):7137–7142
29. Xie G, Luo J, Guo D et al (2010) Nanoconfined ionic liquids under electric fields. Appl Phys Lett 96(4):043112
30. Gallo P, Rovere M, Spohr E (2000) Glass transition and layering effects in confined water: a computer simulation study. J Chem Phys 113(24):11324–11335
31. Hummer G, Rasaiah JC, Noworyta JP (2001) Water conduction through the hydrophobic channel of a carbon nanotube. Nature 414(6860):188–190
32. Koga K, Gao GT, Tanaka H et al (2001) Formation of ordered ice nanotubes inside carbon nanotubes. Nature 412(6849):802–805
33. Takaiwa D, Hatano I, Koga K et al (2008) Phase diagram of water in carbon nanotubes. Proc Natl Acad Sci USA 105(1):39–43
34. Bai J, Wang J, Zeng XC (2006) Multiwalled ice helixes and ice nanotubes. Proc Natl Acad Sci USA 103(52):19664–19667
35. Mochizuki K, Koga K (2015) Solid-liquid critical behavior of water in nanopores. Proc Natl Acad Sci USA 112(27):8221–8226
36. Nomura K, Kaneko T, Bai J et al (2017) Evidence of low-density and high-density liquid phases and isochore end point for water confined to carbon nanotube. Proc Natl Acad Sci USA 114(16):4066–4071
37. Granick S (1999) Soft matter in a tight spot. Phys Today 52(7):26–31
38. Schmidt M, Löwen H (1997) Phase diagram of hard spheres confined between two parallel plates. Phys Rev E 55(6):7228
39. Koga K, Zeng XC, Tanaka H (1997) Freezing of confined water: a bilayer ice phase in hydrophobic nanopores. Phys Rev Lett 79(26):5262
40. Kimmel GA, Matthiesen J, Baer M et al (2009) No confinement needed: observation of a metastable hydrophobic wetting two-layer ice on graphene. J Am Chem Soc 131(35):12838–12844
41. Kimmel GA, Petrik NG, Dohnálek Z et al (2005) Crystalline ice growth on Pt (111): observation of a hydrophobic water monolayer. Phys Rev Lett 95(16):166102
42. Bampoulis P, Teernstra VJ, Lohse D et al (2016) Hydrophobic ice confined between graphene and MoS2. J Phys Chem C 120(47):27079–27084
43. Zangi R, Mark AE (2003) Monolayer ice. Phys Rev Lett 91(2):025502
44. Koga K, Tanaka H (2005) Phase diagram of water between hydrophobic surfaces. J Chem Phys 122(10):104711
45. Bai J, Angell CA, Zeng XC (2010) Guest-free monolayer clathrate and its coexistence with two-dimensional high-density ice. Proc Natl Acad Sci USA 107(13):5718–5722
46. Zhao WH, Wang L, Bai J et al (2014) Highly confined water: two-dimensional ice, amorphous ice, and clathrate hydrates. Acc Chem Res 47(8):2505–2513
47. Zhao WH, Bai J, Yuan LF et al (2014) Ferroelectric hexagonal and rhombic monolayer ice phases. Chem Sci 5(5):1757–1764
48. Kaneko T, Bai J, Yasuoka K et al (2013) New computational approach to determine liquid-solid phase equilibria of water confined to slit nanopores. J Chem Theory Comput 9(8):3299–3310
49. Bai J, Zeng XC (2012) Polymorphism and polyamorphism in bilayer water confined to slit nanopore under high pressure. Proc Natl Acad Sci USA 109(52):21240–21245

50. Bai J, Zeng XC, Koga K et al (2003) Formation of quasi two-dimensional bilayer ice in hydrophobic slits: a possible candidate for ice XIII? Mol Simulat 29(10–11):619–626
51. Han S, Choi MY, Kumar P et al (2010) Phase transitions in confined water nanofilms. Nat Phys 6(9):685–689
52. Giovambattista N, Rossky PJ, Debenedetti PG (2009) Phase transitions induced by nanoconfinement in liquid water. Phys Rev Lett 102(5):050603
53. Jia M, Zhao W, Yuan L (2014) New hexagonal-rhombic trilayer ice structure confined between hydrophobic plates. Chin J Chem Phys 27:15–19
54. Kumar P, Buldyrev SV, Starr FW et al (2005) Thermodynamics, structure, and dynamics of water confined between hydrophobic plates. Phys Rev E 72(5):051503
55. Kastelowitz N, Johnston JC, Molinero V (2010) The anomalously high melting temperature of bilayer ice. J Chem Phys 132(12):124511
56. Zangi R, Mark AE (2004) Electrofreezing of confined water. J Chem Phys 120(15):7123–7130
57. Algara-Siller G, Lehtinen O, Wang FC et al (2015) Square ice in graphene nanocapillaries. Nature 519(7544):443–445
58. London F (1937) The general theory of molecular forces. Trans Faraday Soc 33:8b–26
59. Rowlinson JS, Widom B (2013) Molecular theory of capillarity. Courier Corporation
60. Zhou W, Yin K, Wang C et al (2015) The observation of square ice in graphene questioned. Nature 528(7583):E1–E2
61. Corsetti F, Matthews P, Artacho E (2015) Structural and configurational properties of nanoconfined monolayer ice from first principles. Sci Rep 6:18651
62. Chen J, Schusteritsch G, Pickard CJ et al (2016) Two dimensional ice from first principles: structures and phase transitions. Phys Rev Lett 116(2):025501
63. Chen J, Zen A, Brandenburg JG et al (2016) Evidence for stable square ice from quantum Monte Carlo. Phys Rev B 94(22):220102
64. Strauss I, Chan H, Kral P (2014) Ultralong polarization chains induced by ions solvated in confined water monolayers. J Am Chem Soc 136(4):1170–1173
65. Roman T, Groß A (2016) Polymorphism of water in two dimensions. J Phys Chem C 120(25):13649–13655
66. Corsetti F, Zubeltzu J, Artacho E (2016) Enhanced configurational entropy in high-density nanoconfined bilayer ice. Phys Rev Lett 116(8):085901
67. Satarifard V, Mousaei M, Hadadi F et al (2017) Reversible structural transition in nanoconfined ice. Phys Rev B 95(6):064105
68. Chen J, Schusteritsch G, Pickard CJ et al (2017) Double-layer ice from first principles. Phys Rev B 95(9):094121
69. Zhu W, Zhao WH, Wang L et al (2016) Two-dimensional interlocked pentagonal bilayer ice: how do water molecules form a hydrogen bonding network? Phys Chem Chem Phys 18(21):14216–14221
70. Neek-Amal M, Peeters FM, Grigorieva IV et al (2016) Commensurability effects in viscosity of nanoconfined water. ACS Nano 10(3):3685–3692
71. Mario MSF, Neek-Amal M, Peeters FM (2015) AA-stacked bilayer square ice between graphene layers. Phys Rev B 92(24):245428
72. Fernández MS, Peeters FM, Neek-Amal M (2016) Electric-field-induced structural changes in water confined between two graphene layers. Phys Rev B 94(4):045436
73. Qiu H, Zeng XC, Guo W (2015) Water in inhomogeneous nanoconfinement: coexistence of multilayered liquid and transition to ice nanoribbons. ACS Nano 9(10):9877–9884
74. What don't we know? 125 Science Questions. Science, 2005, 309:78–102
75. Geim AK (2009) Graphene: status and prospects. Science 324(5934):1530–1534
76. Ezawa M (2006) Peculiar width dependence of the electronic properties of carbon nanoribbons. Phys Rev B 73(4):045432
77. Balandin AA, Ghosh S, Bao W et al (2008) Superior thermal conductivity of single-layer graphene. Nano Lett 8(3):902–907
78. Liu M, Yin X, Ulin-Avila E et al (2011) A graphene-based broadband optical modulator. Nature 474(7349):64–67

79. Lee C, Wei X, Kysar JW et al (2008) Measurement of the elastic properties and intrinsic strength of monolayer graphene. Science 321(5887):385–388
80. Stoller MD, Park S, Zhu Y et al (2008) Graphene-based ultracapacitors. Nano Lett 8(10):3498–3502
81. Stankovich S, Dikin DA, Dommett GHB et al (2006) Graphene-based composite materials. Nature 442(7100):282–286
82. Bunch JS, Verbridge SS, Alden JS et al (2008) Impermeable atomic membranes from graphene sheets. Nano Lett 8(8):2458–2462
83. Bao W, Miao F, Chen Z et al (2009) Controlled ripple texturing of suspended graphene and ultrathin graphite membranes. Nat Nanotechnol 4(9):562–566
84. Geim AK, Novoselov KS (2007) The rise of graphene. Nat Mater 6(3):183–191
85. Novoselov KS, Fal VI, Colombo L et al (2012) A roadmap for graphene. Nature 490(7419):192–200
86. Dreyer DR, Park S, Bielawski CW et al (2010) The chemistry of graphene oxide. Chem Soc Rev 39(1):228–240
87. Cheng Q, Duan J, Zhang Q et al (2015) Learning from nature: constructing integrated graphene-based artificial nacre. ACS Nano 9(3):2231–2234
88. Dikin DA, Stankovich S, Zimney EJ et al (2007) Preparation and characterization of graphene oxide paper. Nature 448(7152):457–460
89. Li D, Müller MB, Gilje S et al (2008) Processable aqueous dispersions of graphene nanosheets. Nat Nanotechnol 3(2):101–105
90. Tian Y, Cao Y, Wang Y et al (2013) Realizing ultrahigh modulus and high strength of macroscopic graphene oxide papers through crosslinking of mussel-inspired polymers. Adv Mater 25(21):2980–2983
91. Park S, Lee KS, Bozoklu G et al (2008) Graphene oxide papers modified by divalent ions-enhancing mechanical properties via chemical cross-linking. ACS Nano 2(3):572–578
92. An Z, Compton OC, Putz KW et al (2011) Bio-inspired borate cross-linking in ultra-stiff graphene oxide thin films. Adv Mater 23(33):3842–3846
93. Gao Y, Liu LQ, Zu SZ et al (2011) The effect of interlayer adhesion on the mechanical behaviors of macroscopic graphene oxide papers. ACS Nano 5(3):2134–2141
94. Cheng Q, Wu M, Li M et al (2013) Ultratough artificial nacre based on conjugated cross-linked graphene oxide. Angew Chem 125(13):3838–3843
95. Hu K, Tolentino LS, Kulkarni DD et al (2013) Written-in conductive patterns on robust graphene oxide biopaper by electrochemical microstamping. Angew Chem Inter Ed 52(51):13784–13788
96. Zhang M, Huang L, Chen J et al (2014) Ultratough, ultrastrong, and highly conductive graphene films with arbitrary sizes. Adv Mater 26(45):7588–7592
97. Nair RR, Wu HA, Jayaram PN et al (2012) Unimpeded permeation of water through helium-leak-tight graphene-based membranes. Science 335(6067):442–444
98. Paul DR (2012) Creating new types of carbon-based membranes. Science 335(6067):413–414
99. Li H, Song Z, Zhang X et al (2013) Ultrathin, molecular-sieving graphene oxide membranes for selective hydrogen separation. Science 342(6154):95–98
100. Kim HW, Yoon HW, Yoon SM et al (2013) Selective gas transport through few-layered graphene and graphene oxide membranes. Science 342(6154):91–95
101. Willcox JAL, Kim HJ (2017) Molecular dynamics study of water flow across multiple layers of pristine, oxidized, and mixed regions of graphene oxide. ACS Nano 11(2):2187–2193
102. Radha B, Esfandiar A, Wang FC et al (2016) Molecular transport through capillaries made with atomic-scale precision. Nature 538(7624):222–225
103. Joshi RK, Carbone P, Wang FC et al (2014) Precise and ultrafast molecular sieving through graphene oxide membranes. Science 343(6172):752–754
104. Mi B (2014) Graphene oxide membranes for ionic and molecular sieving. Science 343(6172):740–742
105. Churaev NV, Sobolev VD (1995) Prediction of contact angles on the basis of the Frumkin-Derjaguin approach. Adv Colloid Interf Sci 61:1–16

106. Israelachvili JN (2011) Intermolecular and surface forces. Academic Press
107. Zhao YP (2012) Physical mechanics of surfaces and interfaces. Science Press, Beijing
108. Zhang L, Wang Y, Lv J et al (2017) Materials discovery at high pressures. Nat Rev Mater 2:17005
109. Koga K, Tanaka H, Zeng XC (2000) First-order transition in confined water between high-density liquid and low-density amorphous phases. Nature 408(6812):564–567
110. Kaneko T, Bai J, Akimoto T et al (2018) Phase behaviors of deeply supercooled bilayer water unseen in bulk water. Proc Natl Acad Sci USA 115(19):4839–4844
111. Wang FC, Zhu YB, Wu HA (2018) Structure and transport of confined liquid in nanochannels. SCIENTIA SINICA Physica, Mechanica & Astronomica 48(9):094609

106. Israelachvili JN (2011) Intermolecular and surface forces. Academic Press
107. Zhao YP (2012) Physical mechanics of surfaces and interfaces. Science Press, Beijing
108. Zhang L, Wang Y, Lv J et al (2017) Materials discovery at high pressures. Nat Rev Mater 2:17005
109. Koga K, Tanaka H, Zeng XC (2000) First-order transition in confined water between high-density liquid and low-density amorphous phases. Nature 408(6812):564–567
110. [illegible] E, Bai L, [illegible] T et al (2018) Large population of [illegible] water molecules [illegible] in bulk water. Proc Natl Acad Sci USA 115(19):4848–4853
111. Wang GQ, Zhu YB, Wu HA (2018) Structure and transport of confined water [illegible] (in Chinese). SCIENTIA SINICA Physica, Mechanica & Astronomica 48: [illegible]

Chapter 2
Methods and Models

2.1 Brief Introduction of Water Model

In TIP water models established by Jorgensen et al. [1], O–H bond length and H–O–H bond angle use the experimental measurements, and the main difference among different models is the position of negative charge. In TIP3P water model, the negative charge is located at oxygen atom. In TIP4P model proposed by Bernal and Fowler [2], negative charge is on point M, which located at the angular bisector of H–O–H angle near positive charge and the distance between point M and oxygen atom is defined as d_{OM}. Horn et al. [3] used the Ewald algorithm to optimize the parameters of TIP4P model. Subsequently, Abascal and Vega obtained improved TIP4P/2005 model [4]. In TIP5P model [5], two special charge are located at the positions of lone pair electrons. The geometric feature of TIP5P model is similar to that of ST2 model [6]. Nada and Eerden proposed six-site model [7], which combined the characteristics of four- and five-site models. Whereas, the prediction of melting point based on six-site model is very high [8]. In SPC model proposed by Berendsen et al. [9–11], the O–H bond length is set to be 1.0 Å and the H–O–H bond angle is equal to the angle of tetrahedron. In 1987, Berendsen et al. obtained the SPC/E model [9], which is similar to SPC model while the charge is different. Here, we listed the main parameters of those models in Table 2.1 [12].

Those rigid water models in Table 2.1 exhibit two common features; that is, the LJ interaction site is on oxygen atoms, and position charge is set on hydrogen atoms. SPC, SPC/E, and TIP3P potentials are 2D three-site models. TIP4P, TIP4P/Ew, and TIP4P/2005 potentials are 2D four-site models. TIP5P and TIP5P/Ew are tetrahedral five-site models. Table 2.2 [12] lists the comparisons between experiments and simulations on the melting of bulk ice (I_h) under standard atmospheric pressure. It is found that the results of melting point (T_m) and critical temperature (T_c) obtained from different models exist big differences. Referring to the experimental measurements,

Y. Zhu, *Phase Behavior of Two-Dimensional Water Confined in Graphene Nanocapillaries*, Springer Theses,
https://doi.org/10.1007/978-981-15-7957-8_2

Table 2.1 Parameters of some rigid water models [12]

Model	d_{OH} (Å)	H–O–H (°)	σ (Å)	ε (kJ/mol)	q_H (*e*)	d_{OM} (Å)	d_{OL} (Å)
SPC	1.0	109.47	3.1656	0.6500	0.4100	–	–
SPC/E	1.0	109.47	3.1656	0.6500	0.4238	–	–
TIP3P	0.9572	104.52	3.1506	0.6364	0.4170	–	–
TIP4P	0.9572	104.52	3.1540	0.6480	0.5200	0.1500	–
TIP4P/Ew	0.9572	104.52	3.1643	0.680946	0.52422	0.1250	
TIP4P/2005	0.9572	104.52	3.1589	0.7749	0.5564	0.1546	–
TIP5P	0.9572	104.52	3.1200	0.6694	0.2410	–	0.70
TIP5P/Ew	0.9572	104.52	3.0970	0.7448	0.2410	–	0.70

The d_{OH} is the O–H bond length. The σ and ε are parameters used in Lennard-Jones (LJ) 12–6 potential. The d_{OL} is the distance between oxygen atom and lone pair electrons

Table 2.2 Melting properties of bulk ice obtained from simulations and experiments under the standard atmospheric pressure [12]

Model	SPC	SPC/E	TIP3P	TIP4P	TIP5P	Experiments
T_m (K)	190.5	215	145.6	232.5	273.9	273.15
T_c (K)	593.8	638.6	–	588.2	521.3	647.1
T_m/T_c	0.321	0.337	–	0.394	0.525	0.422
ρ_l (g/cm^3)	0.991	1.011	1.017	1.002	0.987	0.999
ρ_{Ih} (g/cm^3)	0.934	0.950	0.947	0.940	0.967	0.917
H_l (kcal/mol)	−11.64	−12.49	−11.69	−10.98	−10.33	–
H_{Ih} (kcal/mol)	−12.22	−13.23	−11.99	−12.03	−12.08	–
ΔH (kcal/mol)	0.62	0.74	0.3	1.05	1.75	1.44
$d\rho/dT$ (bar/K)	−115	−126	−66	−160	−708	−135

The T_m and T_c denote melting temperature and critical temperature, respectively. The H is enthalpy, and ρ is density. The subscripts l and Ih denote liquid and ice states, respectively

the prediction ability of TIP4P model is better than that of SPC and TIP5P models. Based on the Gibbs ensemble method [13], critical temperature can be obtained from simulations [14, 15]. Table 2.2 also gives the value of T_m/T_c, it is found that the predicted value based on TIP4P model is close to the experimental value, while that of other models is far from the experimental measurement [12].

In 2005, Abascal and Vega obtained the TIP4P/2005 model based on previous TIP4P model [4]. The geometric feature of TIP4P/2005 model is similar to that of TIP4P model, while TIP4P/2005 potential could achieve better fitting on more physical parameters. In Table 2.3, Vega et al. [16] systemically compared and rated the simulation results of several commonly used water models. TIP4P/2005 model achieves the highest score, while the prediction of dielectric constant is poor. The higher quadrupole moment of TIP4P/2005 is benefit to predicting melting point and phase diagram, but it seems responsible for the low dielectric constant. Vega et al.

Table 2.3 Score sheet for several typical water models [16]

Property	Model				
	SPC/E	TIP3P	TIP4P	TIP4P/2005	TIP5P
VLE, T_c	3	1	2	3	0
Surface tension	2	1	2	3	0
ρ ices	2	0	2	3	1
Phase diagram	1	0	2	3	1
T_m melting prop	1	0	1	2.5	2.5
T_{TMD}, α, κ_T	1	0	1	3	2
Structure	2	0	1	2.5	2.5
EOS (high ρ)	3	2	1	3	0
D	3	0	1	3	2
ε_D	3	2	0	1	3
Total Score	21	6	13	27	14

In this table, VLE denotes vapor-liquid equilibria, T_c is critical point, ρ is ice density, T_{TMD} is temperature of maximum density, α and κ_T are thermal coefficients, EOS (high ρ) denotes equation of state of liquid water at high pressure, D is self-diffusion coefficient, and ε_D is dielectric constant

suggested that TIP4P/2005 can be considered as a reliable and cheap model of water offering good performance over a wide range of conditions [16].

The results in Vega et al. [16] demonstrated that TIP4P/2005 model exhibits quite excellent prediction ability on the orthobaric densities, critical temperature, surface tension, densities of different ice phases, phase diagram, melting properties, temperature of maximum density, isothermal compressibility, thermal coefficients and the structure of water and ice covering the temperature range from 120 to 640 K and pressure up to 3GPa [16]. In this thesis, we selected the TIP4P/2005 model as the force field in MD simulations of 2D confined water in graphene nanocapillaries.

2.2 Main Models Used in This Thesis

Graphene nanocapillaries exhibit obvious confinement effect, in which size and pressure are two important factors. Recent experiments indicated that water nanodroplet between two sheets of graphene can transform into monolayer square-like ice under ambient temperature [17], known as the (meta)stable limit of 2D liquid. The formation of monolayer square-like ice is mainly due to the ultrahigh vdW pressure (GPa magnitude) in graphene confinement. Pressure as a basic thermodynamic parameter could influence and change the properties of materials, and even forming new materials through phase transitions [18]. Therefore, phase behavior of 2D water in graphene nanocapillaries is closely related to the pressure in confinement.

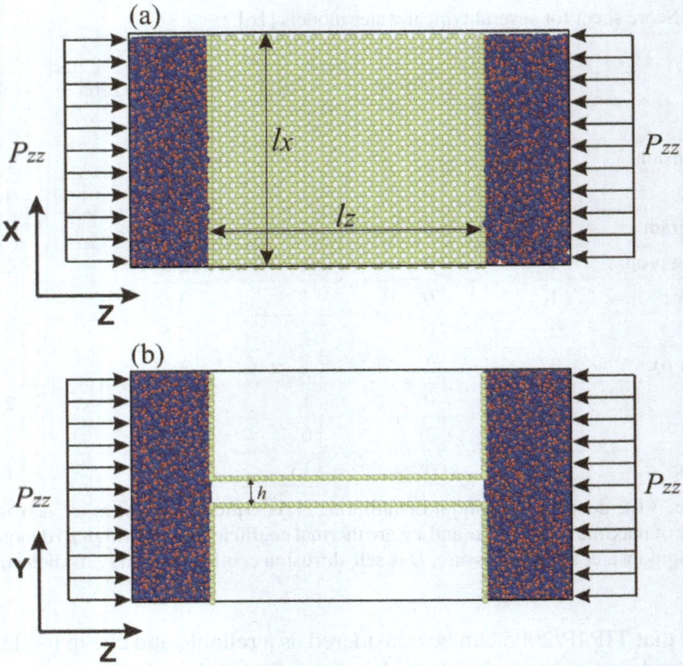

Fig. 2.1 Mechanical model of 2D water confined in graphene nanocapillaries. The channel width *h* is the distance between the center planes of the graphene walls

The main model established in this thesis is shown in Fig. 2.1, in which we used lateral pressure (P_{zz}) to replace the vdW pressure induced by graphene confinement. The simulation system is a graphene nanochannel connected with two water reservoirs. The nanochannel is formed by two sheets of parallel graphene with the center-plane distance of h (defined as the channel width). Each water reservoir contains 1000 water molecules. Water molecules can enter into the graphene channel driven by lateral pressure. Graphene sheets are fixed during the MD simulations. The in-plane size of graphene nanochannel is $l_z \times l_x = 42.60$ Å × 36.89 Å. MD simulations were performed using the LAMMPS program [19]. Periodic boundary conditions were applied on all three directions of the simulation box. The time step was 1 fs. The isothermal-isobaric ensemble ($NP_{zz}T$) was used in simulations, in which the temperature and lateral pressure were controlled by the Nosé-Hoover thermostat and barostat, respectively. We mainly considered three thermodynamic variables: lateral pressure, channel width, and temperature. We could control lateral pressure and channel width to allow one to three monolayers of water to be accommodated within the capillary. The system temperature could also be changed to study the freezing and melting of 2D water/ice in graphene nanocapillaries. In the initial equilibrium run, lateral pressure was kept at 0.1 GPa while temperature was fixed at 300 K for 5 ns, and the water molecules filled the graphene nanocapillary. In MD simulations of compression limit, temperature and channel width were fixed, while lateral pressure increased in a small step of 0.05–0.1 GPa/ns process. In MD simulations, the number of water molecules inside the channel, potential energy (per atom), mean-squared displacement, oxygen-oxygen radial distribution functions, density profiles were outputted to analyze the process of phase transitions. Visualizations were mainly performed using the OVITO program [20].

In MD simulations, the total potential of interaction is taken as the sum of the interactions $\Sigma\phi_{\text{water}}(\mathbf{r}_i,\mathbf{r}_j)$ among water molecules and the external potential $\Sigma\phi_{\text{wall}}(\mathbf{r}_i,\mathbf{r}_j)$ of the interaction between water molecules and the graphene wall, where $\mathbf{r}_i$ is the coordinate of water molecule i, ϕ_{water} is the TIP4P/2005 potential of water, and ϕ_{wall} is the Lennard-Jones (LJ) 12–6 potential for the water-graphene interaction. The pairwise interaction between any two water molecules are described by the extended four-point charge (TIP4P/2005) model [4], including the long-range Coulomb potential and short-range LJ 12–6 potential between the interaction sites. The intermolecular interactions between atom i and j of different molecules are defined by the site-site potential as follows [12]:

$$U_{ij} = 4\varepsilon_{ij}\left[\left(\frac{\sigma_{ij}}{r_{ij}}\right)^{12} - \left(\frac{\sigma_{ij}}{r_{ij}}\right)^{6}\right] + \frac{e^2}{4\pi\varepsilon_0}\frac{q_i q_j}{r_{ij}},$$

where r denotes the distance of separation, q denotes the electric quantity, e is the proton charge, ε_0 is the permittivity of the vacuum, and ε and σ characterize the energy and length scale, respectively. The long-range Coulombic interaction is calculated using a particle-particle particle-mesh solver with an accuracy of 10^{-4}. For the TIP4P/2005 water model [16], the cutoff distance for the LJ interaction is set to be 12.0 and 8.5 Å for the Coulombic interaction. The distance from the oxygen atom to the massless charge is $d_{\text{OM}} = 0.1546$ Å. The LJ potential interaction parameters of water are as follow [12]: $\sigma_{\text{OO}} = 3.1589$ Å, $\varepsilon_{\text{OO}} = 0.1852$ kcal/mol, $\sigma_{\text{OH}} = 0$, $\varepsilon_{\text{OH}} = 0$, $\sigma_{\text{HH}} = 0$, and $\varepsilon_{\text{HH}} = 0$. The carbon-water potentials are based on parameters for carbon, oxygen, and hydrogen obtained through the Lorentz-Berthelot mixing rules [21]:

$$\begin{cases} \varepsilon_{\text{CI}} = \sqrt{\varepsilon_{\text{CC}}\varepsilon_{\text{II}}} \\ \sigma = \frac{1}{2}(\sigma_{\text{CC}} + \sigma_{\text{II}}) \end{cases};\quad \text{I} = \text{O, H}.$$

The carbon-oxygen interaction parameters [21] are $\sigma_{\text{CO}} = 3.19$ Å and $\varepsilon_{\text{CO}} = 0.09369$ kcal/mol. The carbon-carbon interaction parameters are $\sigma_{\text{CC}} = 3.2211$ Å and $\varepsilon_{\text{CC}} = 0.0474$ kcal/mol.

Note that, in recent theoretical simulations, the results from MD simulations based on TIP4P/2005 water model were similar to those from DFT calculations [22–26], such as monolayer square-like ice, monolayer hexagonal ice, bilayer hexagonal ice, and bilayer square-tube ice. We also noted that phase transitions of confined water and ice in most previous studies [27] were based on either four-site TIP4P or the five-site TIP5P water model. Those simulation results are qualitatively the same regardless of the four- or five-site models. The shortcomings of force fields in MD simulations cannot invalidate the results. TIP4P/2005 force field is probably the most accurate of its kind, possibly even representing the best achievable accuracy for a rigid four-site model. The accuracy of DFT calculations is also disputed due to the selected functions. While, current results of 2D confined ice obtained from DFT calculations and MD simulations showed some fine differences. For example, bilayer square-tube ice structures from different simulations displayed different proton arrangements. From the perspective of computational simulations, clearly more studies are needed from the deeper level of theory or the confinement of formation to explore the detail of 2D square ice and the potential phase behavior of 2D water.

Note also that the TIP4P/2005 water model is a rigid model. If the bond angle change is allowed in the TIP4P water model, it is expected that the restriction on the alignment of water molecules can be relaxed to some degree; therefore, the water network could become more close-packed. In any case, the final water network would reflect a subtle interplay among the hydrogen-bonding interaction and packing due to confinement and high lateral pressure, as well as constraint of the intrinsic bond angle of water molecules.

In addition, to simulate the monolayer square-like ice in graphene pocket, we established another model shown in Fig. 2.2. The initial model is two sheets of parallel graphene, in which a nanodroplet of 100 water molecules (in disordered state) was put into the gap. The bottom graphene was fixed as a rigid plane, which the upper graphene could be moved and deformed to adapt the confined water. The size of upper graphene is small than that of bottom graphene. In the perpendicular direction of graphene, both upper and bottom regions were set with 35 Å vacuum

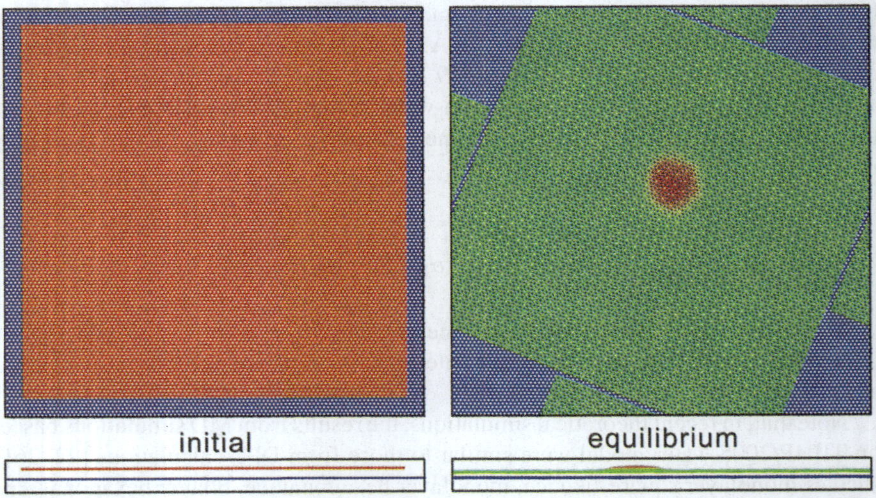

Fig. 2.2 Model of two sheets of graphene. Water nanodroplet is not plotted here. Atoms are colored by height

spaces. The NVT ensemble was used and the temperature was fixed at 300 K. The force field of water used TIP4P/2005 potential. The LJ 12–6 potential was used for graphene-water interaction and graphene-graphene interaction. Due to the strong graphene-graphene interaction, the upper graphene and the bottom graphene sheets come together at margin to form a nanoscale packed pocket which traps the 100 water molecules. After a long time relaxation, the disordered water molecules will spontaneous freeze into an ordered monolayer phase since the high vdW pressure is in the nanoscale confinement of graphene pocket. This model was mainly used to reveal the configurational features of monolayer square-like ice.

2.3 Equilibrium Conditions for Constrained Liquid

Equilibrium conditions for two-phase α and β in the $NP_{zz}T$ ensemble are $T^{\alpha} = T^{\beta}$, $\mu^{\alpha} = \mu^{\beta}$, and ${P_{zz}}^{\alpha} = {P_{zz}}^{\beta}$. In general, ${P_{yy}}^{\alpha} \neq {P_{yy}}^{\beta}$ at equilibrium, where P_{yy} is the pressure perpendicular to the channel plane. We assumed that the area of channel plane is A and channel width is h. The number of water molecules in the channel is N and the entropy is S. The Clapeyron equation for the two-phase boundaries in the quasi-2D constraint system [28]:

$$\begin{cases} \left(\frac{\partial T}{\partial P_{zz}}\right)_h = \frac{v^{\alpha}-v^{\beta}}{s^{\alpha}-s^{\beta}} \\ \left(\frac{\partial P_{zz}}{\partial h}\right)_T = \frac{(a\Delta P)^{\alpha}-(a\Delta P)^{\beta}}{v^{\alpha}-v^{\beta}} \\ \left(\frac{\partial h}{\partial T}\right)_{P_{zz}} = \frac{s^{\alpha}-s^{\beta}}{(a\Delta P)^{\alpha}-(a\Delta P)^{\beta}} \end{cases},$$

where $\Delta P = P_{yy} - P_{zz}$, $s = S/N$, $a = A/N$, and $v = Ah/N$. The Clapeyron equations could be used to check some results obtained from MD simulations.

2.4 Summary

MD simulations play an important role in the studies of confined water, which could investigate and predict the structure, phase transition, transport, and physical and chemical properties of nanoscale confined water under different thermodynamic variables. In this thesis, we mainly concerned the monolayer, bilayer, and trilayer water/ice in graphene nanocapillaries. In the main model (Fig. 2.1), we used mechanical pressure to replace the vdW pressure in graphene confinement, which could well describe the confinement effect of graphene nanocapillaries on 2D confined water. The TIP4P/2005 force field is probably the most accurate of its kind, possibly even representing the best achievable accuracy for a rigid four-site model [16].

Note that, force fields of water molecule remain the problem of precision and some fine differences are existed among different simulation studies, but MD simulation is still one of the most effective research methods in this field. The investigation on confinement pressure-dependent phase behavior of 2D water confined in graphene nanocapillaries is meaningful and significant to theoretically understand low-dimensional water and ice.

References

1. Jorgensen WL, Chandrasekhar J, Madura JD et al (1983) Comparison of simple potential functions for simulating liquid water. J Chem Phys 79(2):926–935
2. Bernal JD, Fowler RH (1933) A theory of water and ionic solution, with particular reference to hydrogen and hydroxyl ions. J Chem Phys 1(8):515–548
3. Horn HW, Swope WC, Pitera JW et al (2004) Development of an improved four-site water model for biomolecular simulations: TIP4P-Ew. J Chem Phys 120(20):9665–9678
4. Abascal JLF, Vega C (2005) A general purpose model for the condensed phases of water: TIP4P/2005. J Chem Phys 123(23):234505
5. Mahoney MW, Jorgensen WL (2000) A five-site model for liquid water and the reproduction of the density anomaly by rigid, nonpolarizable potential functions. J Chem Phys 112(20):8910–8922
6. Stillinger FH, Rahman A (1974) Improved simulation of liquid water by molecular dynamics. J Chem Phys 60(4):1545–1557
7. Nada H, van der Eerden JPJM (2003) An intermolecular potential model for the simulation of ice and water near the melting point: a six-site model of H_2O. J Chem Phys 118(16):7401–7413
8. Abascal JLF, Fernández RG, Vega C et al (2006) The melting temperature of the six site potential model of water. J Chem Phys 125(16):166101
9. Berendsen HJC, Grigera JR, Straatsma TP (1987) The missing term in effective pair potentials. J Phys Chem 91(24):6269–6271
10. Berendsen HJC, Postma JPM, van Gunsteren WF et al (1984) Molecular dynamics with coupling to an external bath. J Chem Phys 81(8):3684–3690
11. Berendsen HJC, Postma JPM, Van Gunsteren WF et al (1981) Intermolecular forces. D. B Pullman, Reidel, Dordrecht
12. Vega C, Sanz E, Abascal JLF (2005) The melting temperature of the most common models of water. J Chem Phys 122(11):114507
13. Panagiotopoulos AZ (1987) Direct determination of phase coexistence properties of fluids by Monte Carlo simulation in a new ensemble. Molecular Phys 61(4):813–826
14. Errington JR, Panagiotopoulos AZ (1998) A fixed point charge model for water optimized to the vapor-liquid coexistence properties. J Phys Chem B 102(38):7470–7475
15. Lísal M, Nezbeda I, Smith WR (2004) Vapor-liquid equilibria in five-site (TIP5P) models of water. J Phys Chem B 108(22):7412–7414
16. Vega C, Abascal JLF, Conde MM et al (2009) What ice can teach us about water interactions: a critical comparison of the performance of different water models. Faraday Discuss 141:251–276
17. Algara-Siller G, Lehtinen O, Wang FC, Nair RR, Kaiser U, Wu HA, Geim AK, Grigorieva IV (2015) Square ice in graphene nanocapillaries. Nature 519:443–445
18. Zhang L, Wang Y, Lv J et al (2017) Materials discovery at high pressures. Nat Rev Mater 2:17005
19. Plimpton S (1995) Fast parallel algorithms for short-range molecular dynamics. J Comput Phys 117(1):1–19
20. Stukowski A (2009) Visualization and analysis of atomistic simulation data with OVITO-the open visualization tool. Model Simul Mater Sci Eng 18(1):015012

21. Werder T, Walther JH, Jaffe RL et al (2003) On the water-carbon interaction for use in molecular dynamics simulations of graphite and carbon nanotubes. J Phys Chem B 107(6):1345–1352
22. Corsetti F, Matthews P, Artacho E (2016) Structural and configurational properties of nanoconfined monolayer ice from first principles. Sci Rep 6:18651
23. Chen J, Schusteritsch G, Pickard CJ, Salzmann CG, Michaelides A (2016) Two dimensional ice from first principles: Structures and phase transitions. Phys Rev Lett 116:025501
24. Roman T, Groß A (2016) Polymorphism of water in two dimensions. J Phys Chem C 120(25):13649–13655
25. Corsetti F, Zubeltzu J, Artacho E (2016) Enhanced configurational entropy in high-density nanoconfined bilayer ice. Phys Rev Lett 116:085901
26. Chen J, Schusteritsch G, Pickard CJ, Salzmann CG, Michaelides A (2017) Double-layer ice from first principles. Phys Rev B 95:094121
27. Zhao WH, Wang L, Bai J, Yuan LF, Yang JL, Zeng XC (2014) Highly confined water: two-dimensional ice, amorphous ice, and clathrate hydrates. Acc Chem Res 47:2505–2513
28. Koga K, Tanaka H (2005) Phase diagram of water between hydrophobic surfaces. J Chem Phys 122:104711

Chapter 3
Monolayer Square-Like Ice Between Two Graphene Sheets

Water constrained between two graphene sheets can form monolayer square-like ice due to the ultrahigh vdW pressure on the order of GPa magnitude. However, the observation of square ice was questioned because of the insufficient experimental interpretation and the slightly rhomboidal feature in simulated monolayer structures. In this chapter, we demonstrate the inherent square characteristic of monolayer square ice from the perspective of structural and dynamic features.

3.1 Brief Introduction

Although omnipresent is interfacial and confined water/ice on Earth, we still known few about the phase behavior of water and ice at low-dimensional scale [1]. Confined water at nanoscale behaves very differently from its bulk counterpart, which can form various low-dimensional crystalline and amorphous ice structures [2–5]. In previous experimental studies, Nair et al. revealed that water should form monolayer solid-like structure in graphene nanocapillaries [6]. Recent experiments demonstrated that vdW pressure can induce water-to-ice transformation in a nanoscale confinement formed by two sheets of graphene [2]. This phase transition is known as the metastability limit of 2D liquid [7]. It is important to explore the formation and structural features of monolayer ice phase in such a nanoscale confinement, which can provide in-depth understanding on the phase behavior of highly constrained water [2]. Those nanoscale confinements are close to the acceptable scale of computer simulations, rendering more practical significance for simulation investigations of 2D confined water in graphene nanocapillaries [8].

In the theoretical studies of quasi-2D ice, MD simulations and DFT calculations are widely used. Monolayer ice was firstly obtained by Zangi and Mark using TIP5P

Y. Zhu, *Phase Behavior of Two-Dimensional Water Confined in Graphene Nanocapillaries*, Springer Theses,
https://doi.org/10.1007/978-981-15-7957-8_3

potential in classical MD simulations [9]. In computational simulations, Zeng and his co-workers found that monolayer ice exists at least six types of crystalline structures [10]; namely, LD-48MI [11], MS-HMI [12], HD-fRMI [12–14], HD-pRMI [12–14], ML-ice V [11, 13–16], and ML-ice VI [10], respectively. Some recent theoretical simulations demonstrated that water confined between two sheets of graphene can spontaneously form various 2D ice phases [17–27]. When the gap between two graphene sheets is about 0.35 nm, the monolayer square ice obtained from simulations is not the strict square lattice. Based on the diffusion Monte Carlo method, Chen et al. [22] gave the evidence that monolayer square ice can be stable under high lateral pressure (2.0 GPa). The results from MD simulations and DFT calculations showed that monolayer square ice exists square and rhombic features simultaneously [7, 17, 18, 22]. In the snapshots of monolayer square ice from MD simulations, the structure is more like a assemble pattern of square and rhombic elements. In the results of DFT calculations, Chen et al. defined the square phase and rhombic phase [17, 22], while the square phase contains rhombic elements and the rhombic phase looks like a polarized water monolayer as reported by Strauss et al. [28]. The 2D square ice phase has become a matter of debate due to the insufficient verification between experiments and simulations. Further theoretical investigations of confined water under high lateral pressure are needed to understand 2D square ice. To the best of our knowledge, a comprehensively simulation study on structural and dynamic characteristics of monolayer square ice with the influence of finite temperature is still lacking.

In this chapter, we discussed the phase transition of monolayer square ice from monolayer liquid water between two sheets of graphene under ambient temperature. From the perspective of structural and dynamic features, we confirmed the monolayer square ice in graphene nanocapillaries.

3.2 Model and Methods

In this chapter, we mainly used the model in Fig. 2.2, in which 100 water molecules are confined in a graphene pocket formed by two sheets of graphene. More details on MD simulations can be found in Sect. 2.2. We also established a small model to obtain the time-averaging results.

3.3 Results and Discussion

3.3.1 Structural Feature in Monolayer Square-Like Ice

The concept of "square ice" had been reported in previous studies on water confined in carbon nanotubes [29–33], in which the *n*-polygonal (n > 3, except helical structure)

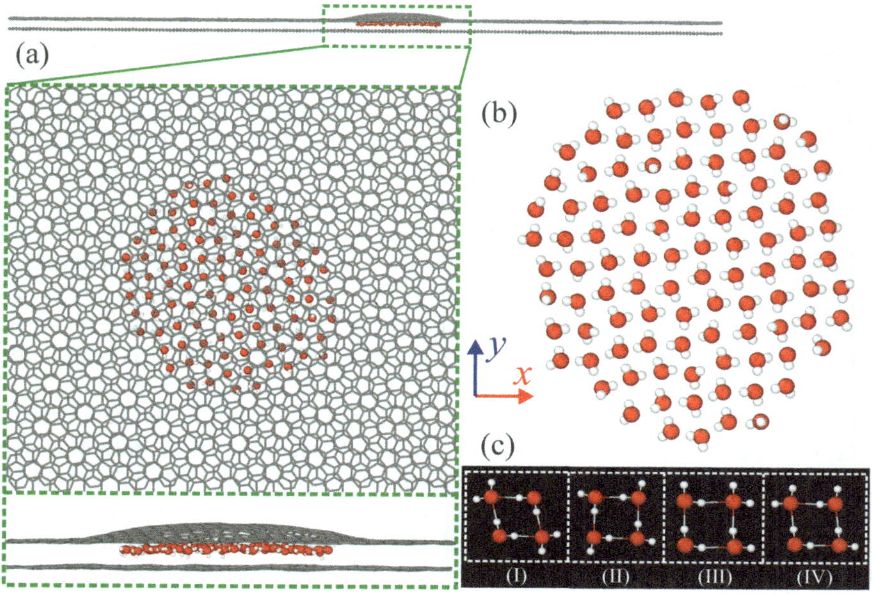

Fig. 3.1 Monolayer square-like ice and its four independent elementary units. **a** The simulation model of 100 water molecules confined in a graphene pocket. Gray network denotes the two sheets of graphene. **b** Typical snapshot of monolayer square-like ice in our MD simulations. **c** Four independent elementary units in monolayer square-like ice

ice nanotubes exhibit a square-net folded by roll-up the vector $(n, 0)$ [31]. Monolayer square-like ice formed between nanochannels was also noted in recent years, while it was usually named as monolayer rhombic ice due to that the hydrogen-bonding network is not a square net strictly [14]. As illustrated in Fig. 3.1, our simulations demonstrate the strong graphene-graphene interaction makes the two graphene sheets come together at the margin and the formed monolayer square-like ice is constrained in a graphene pocket. The structural snapshot in Fig. 3.1b shows that the simulated monolayer square-like ice is slightly rhomboidal, which should be attributed to some rhombic elements and thermal disturbance.

Monolayer square-like ice satisfies the ice rule that every water molecule is hydrogen-bonded to its four nearest neighbors except a few point defects. Based on the hydrogen-bonding network, we find that there are four independent elementary units as shown in Fig. 3.1c, namely, type-I, type-II, type-III, and type-IV, respectively. Both flat and puckered monolayer square-like ices can be assembled by those four units, forming a perfect large square-like hydrogen-bonding network. In the four elementary units, type-II and type-III are square, while the other two are rhombic. In the type-II, the four oxygen atoms form a standard square [22], while the eight hydrogen atoms are not in the same plane. The combination of different units depends on their structural feature and should satisfy the ice rule. In monolayer square-like ice, type-I and type-II units in the whole structure display a long-range ordered pattern with the alternative distribution. Type-IV unit tends to appear in one or more straight columns, while type-III unit is rare. When lateral pressure is higher, more type-IV units would be found in monolayer square-like ice. Note that, the type-IV unit is the same with rhombic phase defined by Chen et al. from DFT calculations [22]. The topologic structure of type-IV units can be seen as a polarized water monolayer [28].

To gain more insight into the structural feature of monolayer square-like ice, we carried out the probability distribution of different elementary units from 10,000 snapshots in MD simulations. Figure 3.2a shows that the probability results of type-I and type-II are nearly equivalent (~43.32% and ~43.38%, respectively) and that of type-IV is ~12.79%, while that of type-III is only ~0.51%. This probability distribution demonstrates that monolayer square-like ice is mainly constituted by type-I and type-II units. The large percent can also indicate that the combined configuration of type-I and type-II units is more stable. In DFT calculations, Corsetti et al. [18] found that the combined configuration of type-I and type-II units exhibits the lowest energy, while the type-III configuration is in a high-energy state. Therefore, monolayer square-like ice obtained from our MD simulations based on TIP4P/2005 model is reasonable and stable. The combined configuration of type-I and type-II units was defined as the square phase by Chen et al. [17], in which the quantum Monte Carlo simulations indicated that this structure is a stable phase in graphene confinement under ~2 GPa lateral pressure [22].

The columns formed by type-IV units in monolayer square-like ice can be seen as stacking fault. Thus, monolayer square-like ice can be defined as assembled square-rhombic network with some stacking faults. As shown in Fig. 3.2b, the arising of stacking faults will change the pattern of hydrogen atoms. In Fig. 3.2b, water molecules with blue protons show regular AB order, while water molecules with green protons display AA order in the perpendicular direction of stacking fault. In MD simulations, we also found the moving of stacking faults in both flat and

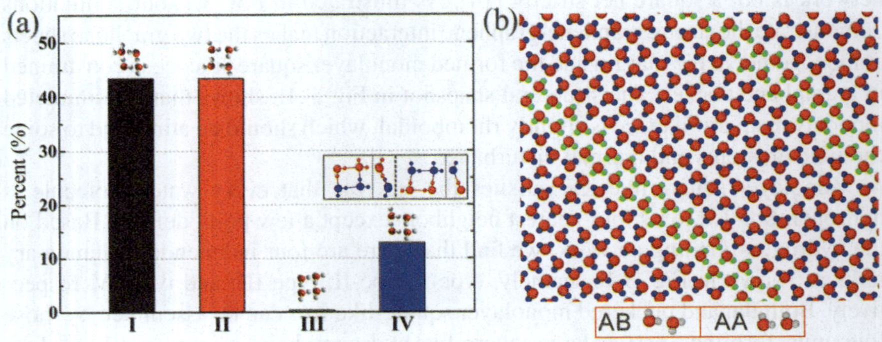

Fig. 3.2 Structural analysis. **a** Probability distribution of the four different elementary units in simulated structures of monolayer square-like ice. **b** Water molecules with blue protons (type-I and type-II) show regular AB order. Water molecules in the stacking faults (polarized water chains) display AA and AB order in two directions

puckered monolayer square-like ice. In previous studies, type-IV units could form stacking faults in the monolayer ice comprising type-III units.

3.3.2 *Spontaneous Flipping Behavior in Monolayer Square-Like Ice*

To further understand the phase behavior of monolayer square-like ice, we analyzed the orientation angle and moving of water molecules in outputted structural snapshots. The stereographic plot of orientation angles (normalized direction vectors) in Fig. 3.3a indicates that most points are located at the edge and those points shows 90° distributions (dashed lines). A few points are far from the edge, indicating that some water molecules are not in the in-plane state. Figure 3.3b shows the orientation angle distribution of water molecules in monolayer square-like ice during 100 ns simulations. Red color is concentrated at the edge, indicating that most of water molecules are in the in-plane state during the 100 ns and the square network is rotated.

In the next, we captured the change of water molecules from the snapshots of monolayer square-like ice during a period of continuous time. In Fig. 3.4, we found the flipping behavior of water molecules in monolayer square-like ice. In Fig. 3.4a, we marked the molecule "**A**" to show the flipping trajectory, which is an obvious out-of-plane flipping [34]. Note that, the MD simulation was performed in NVT ensemble (T = 300 K) without additional constrains and disturbances. Therefore, we could conformed that this flipping behavior is a spontaneous process of water molecules under ambient temperature. In MD simulations, we also found some failed flipping processes. Figure 3.5 illustrates that the flipping of one water molecule could result in the disturbance in around water molecules. This dynamic process would fade away until a metastable configuration formed by the reorientation of surrounding water

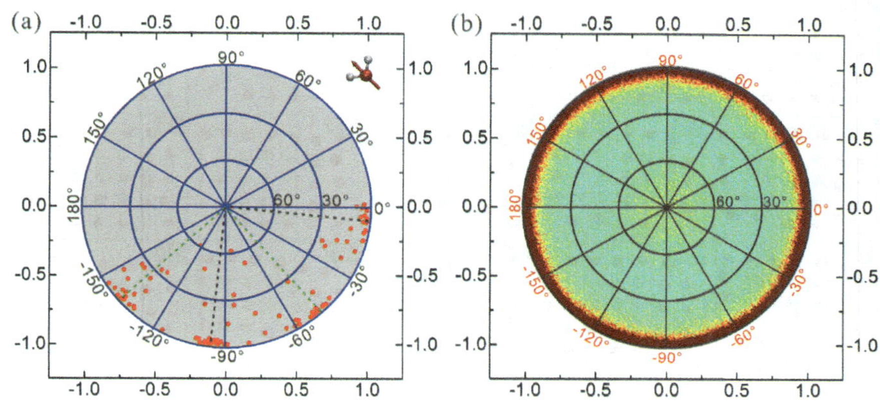

Fig. 3.3 The distributions of the orientation angle of water molecules in monolayer square-like ice. **a** Results of only one snapshot. **b** Results of 100 ns, colored by point density

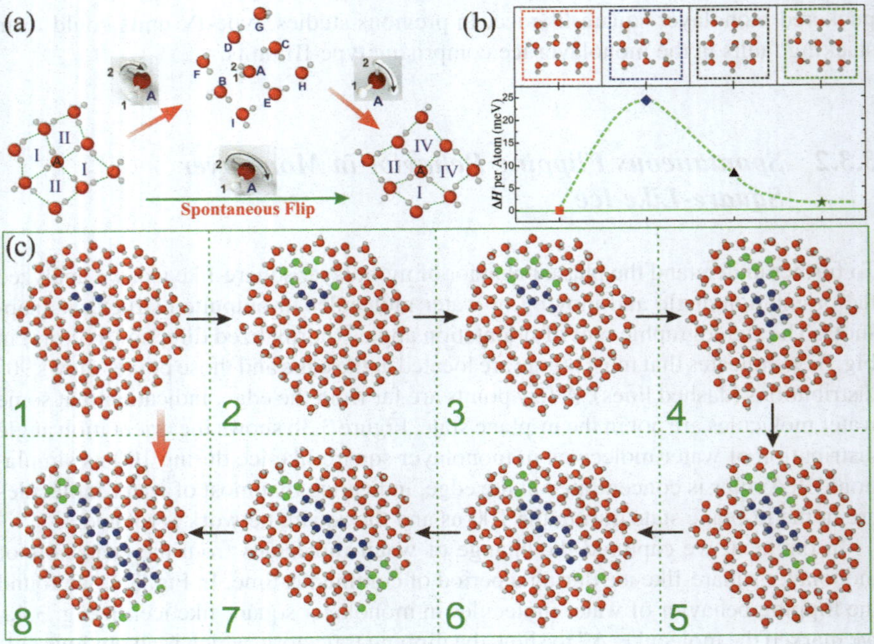

Fig. 3.4 Spontaneous flipping behavior of water molecules in monolayer square-like ice. **a** One captured trajectory of flipping. **b** The change of enthalpy per atom in the flipping. **c** Structural evolution in monolayer square-like ice

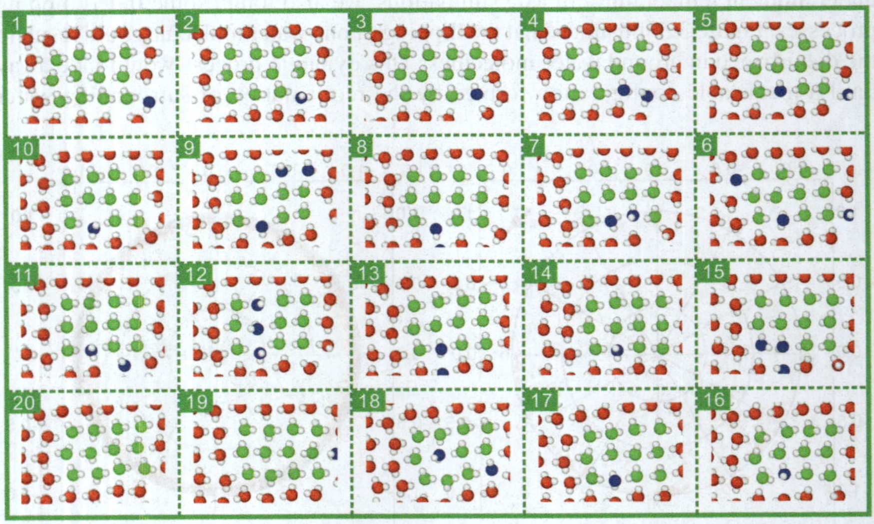

Fig. 3.5 Local structural snapshots of molecular flipping in monolayer square-like ice. Water molecules colored in blue show the flipping process

molecules. The stereographic plot of normalized direction vectors in Fig. 3.3b also shows that flipping behavior could appear in the long-time simulations. Although flipping behavior is common in water/ice system (e.g., the flipping in water chain and various water clusters), the investigations of spontaneous flipping in monolayer square-like ice and associated structural evolution are still lacking.

In Fig. 3.4b, we calculated the change of enthalpy per atom based on the configurations in Fig. 3.4a. The difference of enthalpy (per atom) between initial and final states is ~1.495 meV. Figure 3.4a also show that the flipping will change the type of elementary units. Under the constrain of ice rule, the flipping is accompanied with concerted hydrogen bonds (HBs) breaking and reforming [34, 35]. In graphene nanocapillaries, spontaneous flipping of water molecules is also found in puckered monolayer square-like ice, while that is rarely found in flat monolayer square-like ice due to the too narrow nanochannel. In Figs. 3.4a and 3.5, we could find that the flipping of water molecules would lead to local defects in their instantaneous configurations, and a metastable configuration would be obtained after the reorientation of surrounding water molecules with the change of elementary units. The origination of flipping should be that some weak HBs are broken by thermal disturbance. In monolayer ice, the strength of HBs among different water molecules is not uniform [15]. The existence of weak HBs is mainly due to the unfavorable dipole-dipole interaction between two neighboring straight rows and the thermal disturbance induced position deviations of some water molecules in the tiling network [15]. The breaking of HBs results in local defects, endowing corresponding water molecules with more freedom. Then, the successive flipping would obtain a new metastable structure, while water molecules with failed flipping would come back to their origin states.

To understand the flipping process and associated structural evolution, we captured several successive snapshots of the instantaneous structure from simulated monolayer square-like ice. In Fig. 3.4c, blue and green water molecules are marked to show the flipping trajectory of water molecules. In Fig. 3.4c-2, one blue water appears flipping and several surrounding water molecules (green) show followed slipping. In Fig. 3.4c-3. The seven green water molecules and four blue water molecules finish the flipping process, and another blue water molecule begins the new flipping process. Then, in Fig. 3.4c, 4–7 the flipping process is forward along the trajectory colored by blue. At last, the whole trace is showed in Fig. 3.4c-8. The structural snapshots in Fig. 3.4c indicate that spontaneous flipping of water molecules could spread directionally in the entire structure of monolayer square-like ice. We also found that the type of elementary units near the flipping trace is changed. Two new stacking faults are formed near the flipping trajectory, indicating that the formation of stacking fault in monolayer square-like ice might be related to the flipping behavior of water molecules. The directional flipping trace is mainly due to the formation of new stacking faults. The high-energy state of type-IV units relative to type-I and type-II units results in that the spontaneous flipping trace cannot occur swerve easily to break through the stacking faults (polarized water chains [28]). The directional flipping trace also indicates that the flipping behavior should satisfy the ice rule, instead of irregular change of molecular configuration [14, 34]. This spontaneous slipping behavior should be important for understanding the structural evolution in monolayer square-like ice.

3.3.3 Geometric-Averaged Configuration

The slightly rhomboidal feature in instantaneous snapshots of monolayer square-like ice is mainly due to the existence of type-I and type-IV units. Based on the structural feature in monolayer square-like ice, we found several typical combinations of square units in Fig. 3.6. Here, we mainly considered assembled type-I and type-II units, as well as the type-IV unit. Considering the change of elementary units during the flipping process, we obtained the geometric-averaged configuration through the coordinate conversion of oxygen atoms. In Fig. 3.7a, triangular symbols

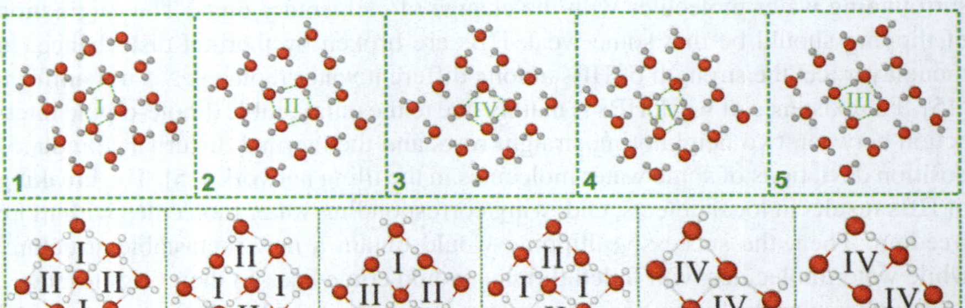

Fig. 3.6 Some typical combinations in monolayer square-like ice

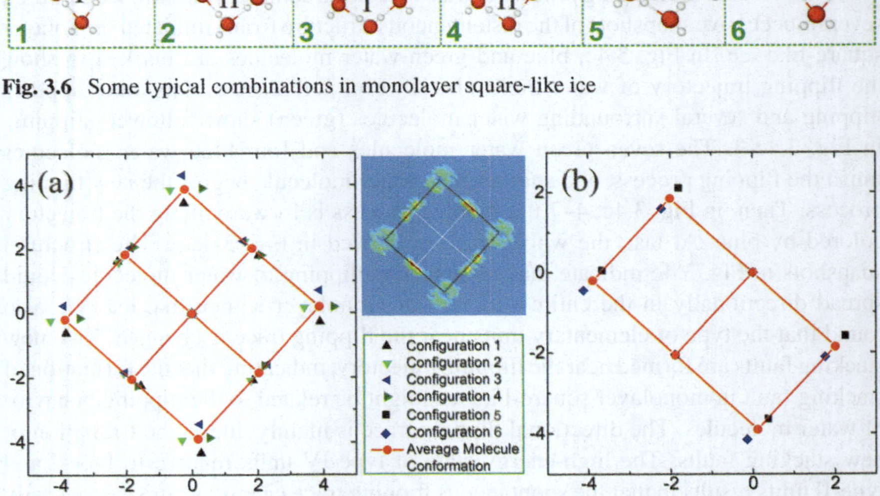

Fig. 3.7 Geometric-averaged configuration through the coordinate conversion of oxygen atoms. The inset between **a** and **b** shows the geometric-averaged configuration of the entire structure of monolayer square-like ice

denote oxygen atoms after coordinate conversion, and circle symbols are the averaged points. We can see that geometric-averaged points can form a perfect square lattice. Figure 3.7b shows the results of type-V units. We also used this method to obtain the geometric-averaged configuration of the entire structure of monolayer square-like ice. As shown in the inset of Fig. 3.7, the configuration (except background) is colored by the density of coordinate points, and black and white lines form the square lattice.

3.3.4 Time Average

Based on the spontaneous flipping behavior and geometric-averaged configuration, we could deduce that the square pattern observed in experiments should be also existed in our long-time MD simulations. To confirm this time averaging results, we carried out MD simulations of 80 water molecules confined between two sheets of graphene with periodic boundary conditions in the two in-plane directions. The channel width is 7.0 Å and temperature is $T = 300$ K. In the model, 80 disordered water molecules were put into the graphene channel. Initial monolayer square-like ice structure was obtained after long-time relaxing within NVT ensemble and energy minimization. The initial structure of monolayer square-like ice is only composed of type-I and type-II units. To observe the flipping and configuration evolution more conveniently, artificial flipping on one water molecule (colored in green in Fig. 3.8) was randomly introduced to destroy the initial equilibrium configuration.

As shown in Fig. 3.8, the randomly introduced artificial flipping would result in the evolution of the entire monolayer structure. The green water molecule would firstly lead to the development of flipping along a trace with the forming of type-IV units. Then, this metastable structure in Fig. 3.8b would appear entire flipping during the long-time simulation. At last, the final configuration in Fig. 3.8c is similar with the initial perfect structure, while each water molecule is flipped and the type-I

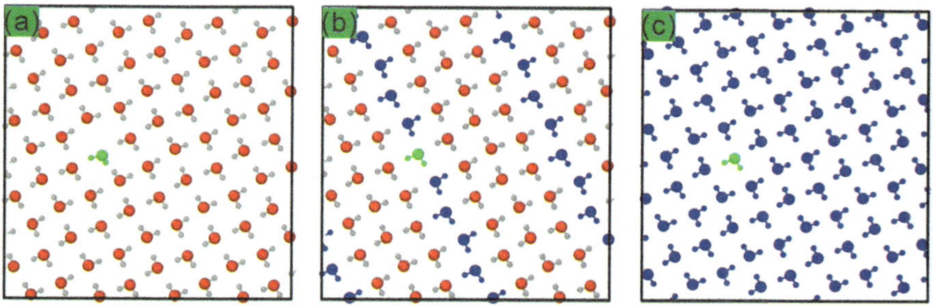

Fig. 3.8 MD simulations of structural flipping through the artificial flipping of one water molecule. Water molecules colored in green show the artificial flipping. Water molecules colored in blue show the flipping of entire structure

and type-II units are exchanged. Figure 3.10a plots the change of potential energy (per atom) during the MD simulation. The combined configuration of type-I and type-II units exhibits the lowest potential energy of −3.603 kcal/mol. During the flipping, the potential energy of metastable structure is -3.570 kcal/mol. Before the entire flipping, the potential energy of the unstable structure is -3.508 kcal/mol. The inset in Fig. 3.10a shows the activation energy and energy difference based on the change of potential energy. The activation energy is 0.062 kcal/mol, and the energy difference between metastable and final structures is 0.033 kcal/mol. Thus, the effective temperature ($\Delta T = \Delta E/(k_B N_A)$, where k_B and N_A are Boltzmann and Avogadro constants, respectively) is 16.60 K. This small value of the effective temperature indicates that flipping of water molecules in monolayer square-like ice can easily occur under ambient temperature.

Moreover, geometric-averaged square pattern can also be obtained from the model of 80 water molecules. We used the snapshots after energy minimization to perform coordinate conversion. The square lattice plotted in Fig. 3.10b illustrates the inherent square characteristic on geometry. Eight neighbor molecules in the vicinity of each water molecule are located on two white dashed circles with the radius of ~2.78 Å and ~3.95 Å, respectively. The centers of the eight red locations and the white point in Fig. 3.10b form four strict squares. The first-neighbor distance between oxygen atoms in Fig. 3.10b is ~2.78 Å, which is similar with the first-peak location (2.79 ± 0.06 Å) in oxygen-oxygen radial distribution function [7] and the measured value (~2.83 Å) in the experiments [2]. To obtain the time-averaged structural evolution, the random velocity excitation was applied on one randomly selected water molecule and then the system is relaxed in the *NVT* ensemble for 10 ns to ensure complete flipping of entire structure. In our simulations, totally 100 times random excitations were applied to observe the long-time-averaging results. The time-averaged results are shown in Figs. 3.9 and 3.10c. The color in Fig. 3.10a denotes the intensive degree of oxygen atoms (coordinate points). The center distance of two adjacent red islands is 2.80 ± 0.05 Å. Based on the results of configurations and dynamic characteristic, we find that the slightly rhomboidal lattice of monolayer square-like ice in long-time MD simulations can average out into one that is indistinguishable from a perfect square pattern.

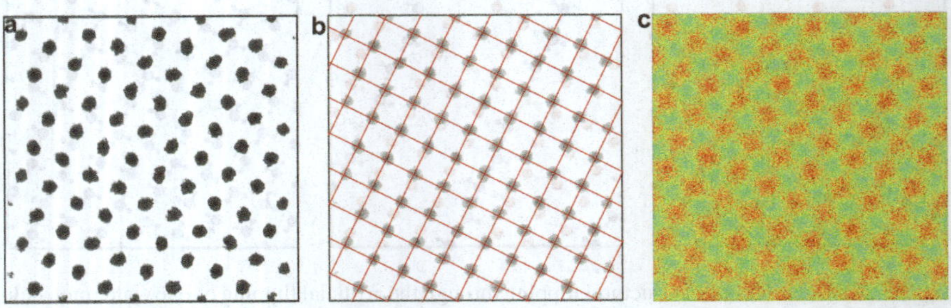

Fig. 3.9 Time averaging square pattern obtained from multiple times random flipping of water molecules during long-time simulations

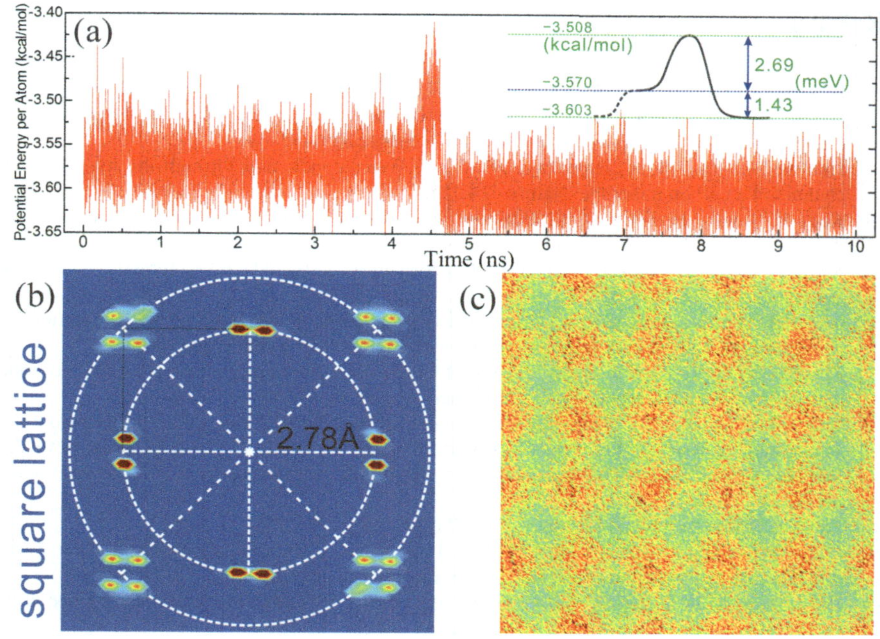

Fig. 3.10 Square pattern of monolayer square-like ice obtained from MD simulations. **a** Potential energy per atom. **b** Configurational-averaged lattice. **c** Time-averaged square lattice selected from Fig. 3.9c

3.4 Summary

In monolayer square-like ice obtained from MD simulations, square and rhombic units are assembled regularly in the order of alternative distribution, while the other rhombic unit forms stacking faults (polarized water chains). Thermal disturbance under ambient temperature would induce the spontaneous flipping of water molecules, resulting in the transformation among different elementary units and the entire structural evolution of monolayer square-like ice. The formation of stacking faults should be related to the flipping behavior of water molecules in square phase. The inherent square pattern of monolayer square-like ice in MD simulations is demonstrated by statistical averaging methods with geometry- and time-averaging effects. We find that the slightly rhomboidal lattice of monolayer square-like ice in long-time MD simulations can average out into one that is indistinguishable from a perfect square pattern. (The investigation in this chapter has been published in *J. Chem. Phys.*, 2017, **147**, 044706.)

References

1. Brovchenko I, Oleinikova A (2008) Interfacial and confined water. Elsevier, Amsterdam
2. Algara-Siller G, Lehtinen O, Wang FC, Nair RR, Kaiser U, Wu HA, Geim AK, Grigorieva IV (2015) Square ice in graphene nanocapillaries. Nature 519:443–445
3. Hummer G, Rasaiah JC, Noworyta JP (2001) Water conduction through the hydrophobic channel of a carbon nanotube. Nature 414:188
4. Koga K, Tanaka H, Zeng XC (2000) First order transition in confined water between high-density liquid and low-density amorphous phases. Nature 408:564–567
5. Koga K, Gao GT, Tanaka H, Zeng XC (2001) Formation of ordered ice nanotubes inside carbon nanotubes. Nature 412:802
6. Nair RR, Wu HA, Jayaram PN, Grigorieva IV, Geim AK (2012) Unimpeded permeation of water through helium-leak-tight graphene-based membranes. Science 335:442
7. Zhu YB, Wang FC, Bai J, Zeng XC, Wu HA (2015) Compression limit of two-dimensional water constrained in graphene nanocapillaries. ACS Nano 9:12197–12204
8. Radha B, Esfandiar A, Wang FC, Rooney AP, Gopinadhan K, Keerthi A, Mishchenko A, Janardanan A, Blake P, Fumagalli L, Lozada-Hidalgo M, Garaj S, Haigh SJ, Grigorieva IV, Wu HA, Geim AK (2016) Molecular transport through capillaries made with atomic-scale precision. Nature 538:222
9. Zangi R, Mark AE (2003) Monolayer ice. Phys Rev Lett 91:025502
10. Zhao WH, Wang L, Bai J, Yuan LF, Yang JL, Zeng XC (2014) Highly confined water: two-dimensional ice, amorphous ice, and clathrate hydrates. Acc Chem Res 47:2505–2513
11. Bai J, Angell CA, Zeng XC (2010) Guest-free monolayer clathrate and its coexistence with two-dimensional high-density ice. Proc Natl Acad Sci USA 107:5718
12. Zhao WH, Bai J, Yuan LF, Yang J, Zeng XC (2014) Ferroelectric hexagonal and rhombic monolayer ice phases. Chem Sci 5:1757–1764
13. Kaneko T, Bai J, Yasuoka K, Mitsutake A, Zeng XC (2013) New computational approach to determine liquid-solid phase equilibria of water confined to slit nanopores. J Chem Theory Comput 9:3299
14. Kaneko T, Bai J, Yasuoka K, Mitsutake A, Zeng XC (2014) Liquid-solid and solid-solid phase transition of monolayer water: High-density rhombic monolayer ice. J Chem Phys 140:184507
15. Koga K, Tanaka H (2005) Phase diagram of water between hydrophobic surfaces. J Chem Phys 122:104711
16. Qiu H, Zeng XC, Guo W (2015) Water in inhomogeneous nanoconfinement: coexistence of multilayered liquid and transition to ice nanoribbons. ACS Nano 9:9877–9884
17. Chen J, Schusteritsch G, Pickard CJ, Salzmann CG, Michaelides A (2016) Two dimensional ice from first principles: Structures and phase transitions. Phys Rev Lett 116:025501
18. Corsetti F, Matthews P, Artacho E (2016) Structural and configurational properties of nanoconfined monolayer ice from first principles. Sci Rep 6:18651
19. Corsetti F, Zubeltzu J, Artacho E (2016) Enhanced configurational entropy in high-density nanoconfined bilayer ice. Phys Rev Lett 116:085901
20. Mario MSF, Neek-Amal M, Peeters FM (2015) AA-stacked bilayer square ice between graphene layers. Phys Rev B 92:245428
21. Neek-Amal M, Peeters FM, Grigorieva IV, Geim AK (2016) Commensurability effects in viscosity of nanoconfined water. ACS Nano 10:3685
22. Chen J, Zen A, Brandenburg JG, Alfe D, Michaelides A (2016) Evidence for stable square ice from quantum Monte Carlo. Phys Rev B 94:220102
23. Zhu YB, Wang FC, Wu HA (2016) Superheating of monolayer ice in graphene nanocapillaries. J Chem Phys 146:134703
24. Chen J, Schusteritsch G, Pickard CJ, Salzmann CG, Michaelides A (2017) Double-layer ice from first principles. Phys Rev B 95:094121
25. Zhu YB, Wang FC, Bai J, Zeng XC, Wu HA (2016) Formation of trilayer ices in graphene nanocapillaries under high lateral pressure. J Phys Chem C 120:8109–8115

26. Zhu YB, Wang FC, Wu HA (2016) Buckling failure of square ice-nanotube arrays constrained in graphene nanocapillaries. J Chem Phys 145:054704
27. Zhu YB, Wang FC, Bai J, Zeng XC, Wu HA (2016) AB-stacked square-like bilayer ice in graphene nanocapillaries. Phys Chem Chem Phys 18:22039–22046
28. Strauss I, Chan H, Kral P (2014) Ultralong polarization chains induced by ions solvated in confined water monolayers. J Am Chem Soc 136:1170–1173
29. Kolesnikov AI, Zanotti JM, Loong CK, Thiyagarajan P, Moravsky AP, Loutfy RO, Burnham CJ (2004) Anomalously soft dynamics of water in a nanotube: a revelation of nanoscale confinement. Phys Rev Lett 93:035503
30. Bai J, Wang J, Zeng XC (2006) Multiwalled ice helixes and ice nanotubes. Proc Natl Acad Sci USA 103:19664
31. Takaiwa D, Hatano I, Koga K, Tanaka H (2008) Phase diagram of water n carbon nanotubes. Proc Natl Acad Sci USA 105:39
32. Tanaka H, Koga K (2005) Formation of ice nanotube with hydrophobic guests inside carbon nanotube. J Chem Phys 123:094706
33. Mochizuki K, Koga K (2015) Solid-liquid critical behavior of water in nanopores. Proc Natl Acad Sci USA 112:8221
34. Richardson JO, Perez C, Lobsiger S, Reid AA, Temelso B, Shields C, Kisiel Z, Wales DJ, Pate BH, Althorpe SC (2016) Concerted hydrogen-bond breaking by quantum tunneling in the water hexamer prism. Science 351:1310
35. Sugimoto T, Aiga N, Otsuki Y, Watanabe K, Matsumoto Y (2016) Emergent high-Tc ferroelectric ordering of strongly correlated and frustrated protons in a heteroepitaxial ice film. Nat Phys 12:1063

Chapter 4
Superheating Behavior of Monolayer Ice in Graphene Nanocapillaries

In the heating process, monolayer ice will transform into liquid phase, while this melting behavior under high lateral pressure is seldom concerned. This chapter focuses on the superheating of monolayer ice in graphene nanocapillaries. We observed four different melting scenarios and monolayer triangular amorphous ice from MD simulations.

4.1 Brief Introduction

Confined water at nanoscale could transform into low-dimensional crystalline or amorphous ice phases under ambient temperature and pressure [1]. Their melting points might extremely increase with reducing the dimensions of confinements compared to bulk ice [2]. Phase transitions among different low-dimensional ice phases are very sensitive to the temperature, pressure, confinement geometry and other unknown factors. A better understanding of the effects of lateral pressure, temperature, and confinement variables on the phase behavior of low-dimensional water/ice would further the advancement of scientific researches and technological applications [3].

The freezing and melting of low-dimensional water/ice are still open questions and remain a matter of debate [2]. Liquid-to-solid, solid-to-solid, and solid-to-liquid-to-solid phase transitions have been found in MD simulations of confined water, including first-order and continuous-like phase transitions [3]. The first-order phase transition between the high-density liquid phase and low-density amorphous phase was firstly found by Koga et al. in MD simulations [4]. Zangi and Mark observed the sudden change in potential energy and the coexistence of two different phases during the melting of monolayer ice, indicating that the melting of monolayer ice is a

Y. Zhu, *Phase Behavior of Two-Dimensional Water Confined in Graphene Nanocapillaries*, Springer Theses,
https://doi.org/10.1007/978-981-15-7957-8_4

first-order phase transition [5]. Bai and Zeng observed an arisen intermediate liquid state (Oswald staging phenomenon) in the phase transition from bilayer hexagonal ice and bilayer square-tube ice [6]. Han et al. revealed that the freezing of bilayer water depends on the density [7]. Two different phase transitions were found in their MD simulations when channel width is 8 Å; that is, the first-order phase transition between low-density liquid and bilayer hexagonal ice and the continuous-like phase transition between high-density liquid and bilayer rhombic amorphous ice [7], respectively. Under ambient temperature, when liquid water is constrained in graphene nanocapillaries, the formation of monolayer square-like ice is a first-order phase transition, while that of bilayer triangular ice is a continuous-like phase transition [8–10]. Our recent MD simulations also revealed that lateral pressure can be as a controlling parameter to realize the first-order and continuous-like phase transitions of 2D confined water [8]. Although 2D square ice in experiments has been a matter of debate [11], the concept of ultrahigh vdW pressure in graphene nanocapillaries has been accepted by researchers. Recent MD simulations and DFT calculations revealed that water confined in graphene nanocapillaries can spontaneously form various 2D crystalline and amorphous ice phases [8–22]. However, the comprehensive investigation on the superheating behavior of monolayer ice in graphene nanocapillaries is still lacking. It is necessary to understand the phase behavior and determine the order of transition in the superheating limit of monolayer ice in graphene nanocapillaries.

In this chapter, we performed MD simulations to study the phase transitions during the superheating process of monolayer ice in graphene nanocapillaries. We found that the superheating of monolayer ice under ultrahigh lateral pressure is distinctly different from the traditional solid-to-liquid melting transition. Monolayer triangular amorphous ice was obtained in our MD simulations. We observed four different superheating scenarios and associated superheating phase diagram of monolayer ice.

4.2 Model and Methods

In this chapter, we used the main model illustrated in Fig. 2.1 to investigate the superheating of monolayer ice in graphene nanocapillaries. The channel width (*h*) was set to be 6.0, 6.5, 7.0 Å, allowing monolayer water monolayer accommodated within the channel. More details on MD simulations can be found in Sect. 2.2. We could change the temperature and lateral pressure to observe the phase transitions of 2D water in graphene nanocapillaries.

4.3 Results and Discussion

4.3.1 Structures of Monolayer Ice in Graphene Nanocapillaries

When lateral pressure is larger than the corresponding compression limit, monolayer water inside graphene nanocapillaries can transform into monolayer square-like ice [8]. The flat square-like hydrogen-bonding network will be puckered when the channel width increases slightly [3]. Figure 4.1 shows three different monolayer ice structures under ambient temperature; that is, flat monolayer square-like ice (fMSI, $h = 6.5$ Å, $P_{zz} = 1.0$ GPa), puckered monolayer square-like ice (pMSI, $h = 7.0$ Å, $P_{zz} = 1.0$ GPa), and highly puckered monolayer square-like ice (hpMSI, $h = 7.0$ Å, $P_{zz} = 2.0$ GPa), respectively. From the top view of hpMSI, the hydrogen-bonding network shows obvious structural features of rectangle and rhombus due to the ultra-high lateral pressure. Especially, we also found monolayer triangular amorphous ice (ML-T) under high temperature and high lateral pressure, as shown in Fig. 4.2. Monolayer triangular phase had been found by Schmidt and Löwen using the model of hard spheres between two parallel planes [23, 24]. ML-T is an amorphous phase, in which oxygen atoms display ordered triangular pattern while the proton arrangement is disorder. The configuration of water molecules in ML-T is similar to that in bilayer AA-stacked triangular ice. The evenly spaced green dashed lines in Fig. 4.2a display the regular triangular pattern. The oxygen-oxygen radial distribution function [$g_{O\text{-}O}(r)$] in Fig. 4.2b demonstrates that the distribution of oxygen atoms in ML-T is regular and the first-neighbor distance of oxygen atoms is 2.80 ± 0.1 Å. The hydrogen-bonding networks of fMSI, pMSI, and hpMSI satisfy the ice rule except few point defects [25, 26]. While, due to the disorder proton pattern, ML-T cannot form a complete hydrogen-bonding network.

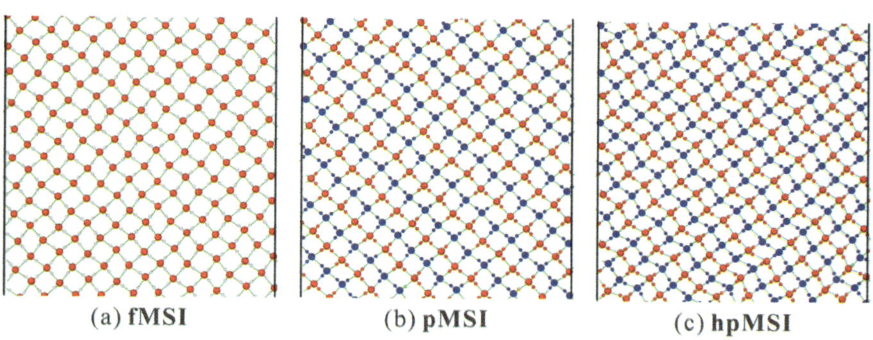

Fig. 4.1 Three different structures of monolayer ice under ambient temperature. **a** Snapshot of flat monolayer square-like ice (fMSI). **b** Snapshot of puckered monolayer square-like ice (pMSI). **c** Snapshot of highly puckered monolayer square-like ice (hpMSI). Water molecules in **b** and **c** are colored by the height in the direction of channel width

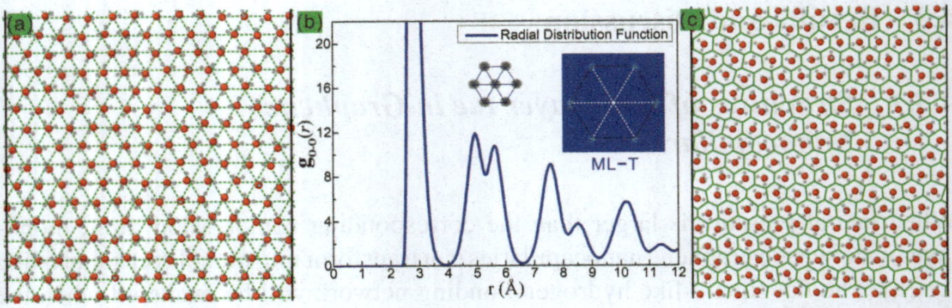

Fig. 4.2 Structure of monolayer triangular amorphous ice (ML-T). **a** Snapshot of ML-T (after energy minimization). The triangular pattern is plotted by equidistant dashed lines. **b** Oxygen-oxygen radial distribution function. **c** Voronoi tessellation diagram of one snapshot of ML-T

To shed light on the structural feature of ML-T, we analyzed the nearest-neighbor positional distribution of each oxygen atom in ML-T during 1 ns equilibrium run. The inset in Fig. 4.2b shows that each water molecule in ML-T has six neighbor molecules which form a hexagon. Voronoi tessellation diagram is plotted in Fig. 4.2c based on the positions of oxygen atoms in ML-T, in which most of polygons are hexagon. The structural snapshots of ML-T demonstrate that some hydrogen atoms are in out-of-plane state due to the unstable hydrogen-bonding network. It could be deduced that ultrahigh lateral pressure and high temperature result in the destruction of hydrogen-bonding network in monolayer square-like ice. The mismatched hydrogen bonds lead to the sixfold orientational order in ML-T. Note that, ML-T herein is different from the triangular phase predicted by Roman and Groß [27]. The triangular phase obtained from DFT calculations is a rhombic structure with ordered proton pattern [27], while ML-T is an amorphous phase transformed from monolayer square-like ice during the superheating process under ultrahigh lateral pressure.

Under different conditions of lateral pressure ($P_{zz} = 1.0$, 2.0, and 3.0 GPa), temperature ($T = 300$, 400, 500 K), and channel width ($h = 6.0$, 6.5, and 7.0 Å), the oxygen density profiles of monolayer water confined in graphene nanocpaillaries are plotted in Fig. 4.3. Under ambient temperature, the formations of fMSI and pMSI depend on the channel width. When the channel width is 6.0 Å, oxygen density profiles

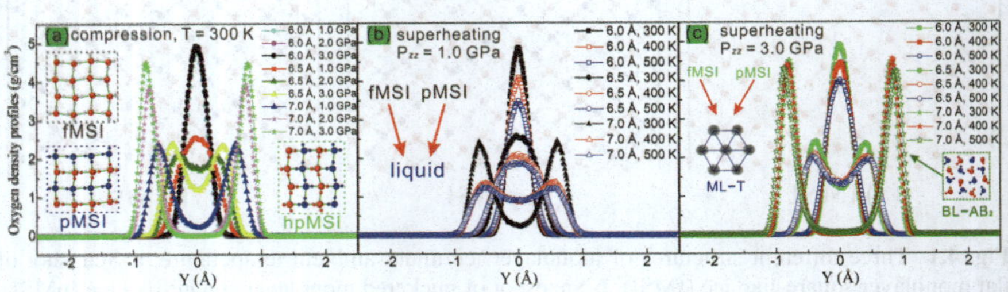

Fig. 4.3 Density profiles of oxygen atoms of monolayer water confined in graphene nanocapillaries under different conditions of temperature, channel width and lateral pressure

remain almost the same under different lateral pressure. Under ultrahigh lateral pressure, the oxygen density profile of hpMSI exhibits two obviously independent peaks. During the superheating process, oxygen density profiles manifest distinctly different changes dependent to lateral pressure. When the lateral pressure is 1.0 GPa, the peak values of oxygen density profiles under higher temperature show obvious reduction, indicating the phase transitions from fMSI and pMSI to liquid state. While under ultrahigh lateral pressure (3.0 GPa), the change on oxygen density profiles is not obvious. When the channel width is 6.0 or 6.5 Å, ML-T is formed during the superheating process. when the channel width is 7.0 Å, a bilayer AB-stacked amorphous ice is formed under high temperature. The results in Fig. 4.3 indicate that the superheating behavior of monolayer ice in graphene nanocapillaries strongly depends on the lateral pressure and channel width. Therefore, it is full of necessity to understand the superheating (melting) of monolayer ice in graphene nanocapillaries.

4.3.2 *Melting of Monolayer Square-Like Ice Under Low Lateral Pressure*

Based on the differences among oxygen density profiles in Fig. 4.3, we performed extensive MD simulations to investigate the superheating behavior of monolayer ice in graphene nanocapillaries. At first, we investigated the melting process of fMSI and pMSI under low lateral pressure ($P_{zz} = 1.0$ GPa).

Figure 4.4 illustrates the independent MD simulations of fMSI ($h = 6.0$ Å) and pMSI ($h = 6.5$ Å) during heating and cooling processes. In the heating process, potential energy (per molecule) first increases gradually and then suddenly jumps by about 0.2 kcal/mol. This rapid change in potential energy is due to the solid-to-liquid phase transition in the superheating process. When the temperature is beyond the superheating limit, the ordered hydrogen-bonding networks in fMSI and pMSI are destroyed. In the reverse cooling process, potential energy decreases gradually and then drops suddenly, overlapping with the most path of heating process. The sudden drop in potential energy during the freezing process is due to the liquid-to-solid phase transition. Figure 4.4 also gives the oxygen-oxygen radial distribution functions at different temperature, which can reflect the change between ordered and disordered phases. The obvious hysteresis loop in potential energy demonstrate that the phase transition between monolayer square-like ice and monolayer liquid water is a first-order phase transition [4, 8]. Especially, the heating/cooling processes trigger the melting/freezing transitions, giving the melting/freezing points of 2D monolayer square-like ice. Therefore, under low lateral pressure ($P_{zz} = 1.0$ GPa), the simulation results in Fig. 4.4 reveal that the melting of monolayer square-like ice is a first-order phase transition [5].

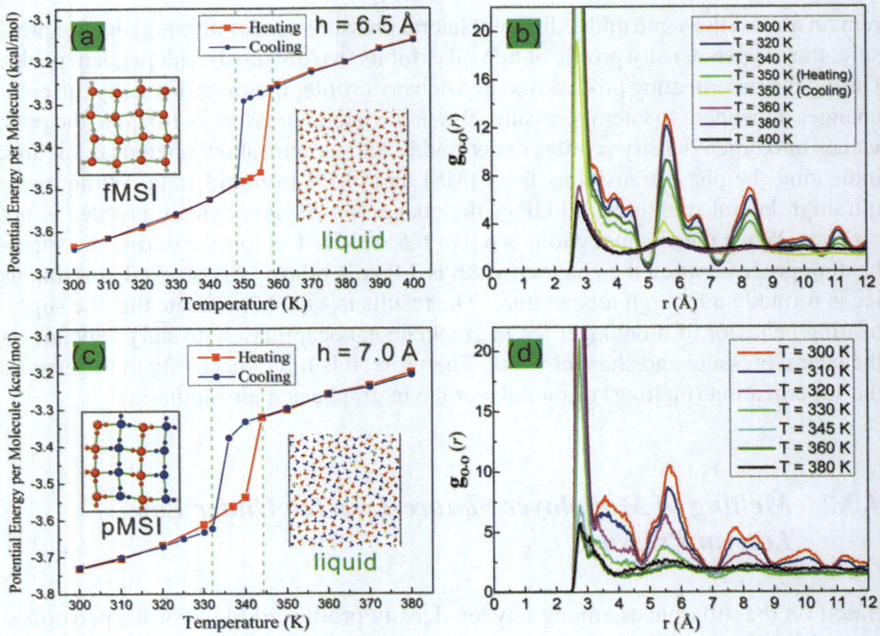

Fig. 4.4 First-order melting phase transitions of fMSI ($h = 6.0$ Å) and pMSI ($h = 6.5$ Å) when the lateral pressure is fixed at 1.0 GPa

4.3.3 *Melting of Monolayer Square-Like Ice Under Ultrahigh Lateral Pressure*

We noted that previous studies on the melting of low-dimensional ice seldom considered the effect of high lateral pressure. Recent MD simulations and DFT calculations revealed that high lateral pressure can result in the formation of various 2D crystalline and amorphous ice phases in graphene nanocapillaries [8–22, 27]. While, under high compression pressure and high degree of superheating, the metastability limit extension of phase diagram of 2D liquid water is yet to be studied in detail. In our MD simulations, we found that the intermediate ML-T phase will be formed during the superheating process of both fMSI and pMSI under ultrahigh lateral pressure, which is distinctly different from first-order phase transition under low lateral pressure.

In Fig. 4.5a, the potential energy exhibits a sudden increase at $T = $ ~412 K, indicating a first-order phase transition from fMSI to ML-T ($h = 6.0$ Å and $P_{zz} = 3.5$ GPa). This rapid change in potential energy is due to the destroy of square-like hydrogen-bonding network during the superheating process. To confirm the phase transition from fMSI to ML-T, we carried out independent MD simulations of heating and cooling processes (350 K $\leq T \leq$ 450 K). In Fig. 4.6a, the hysteresis loop in potential energy indicates the first-order phase transition between fMSI and

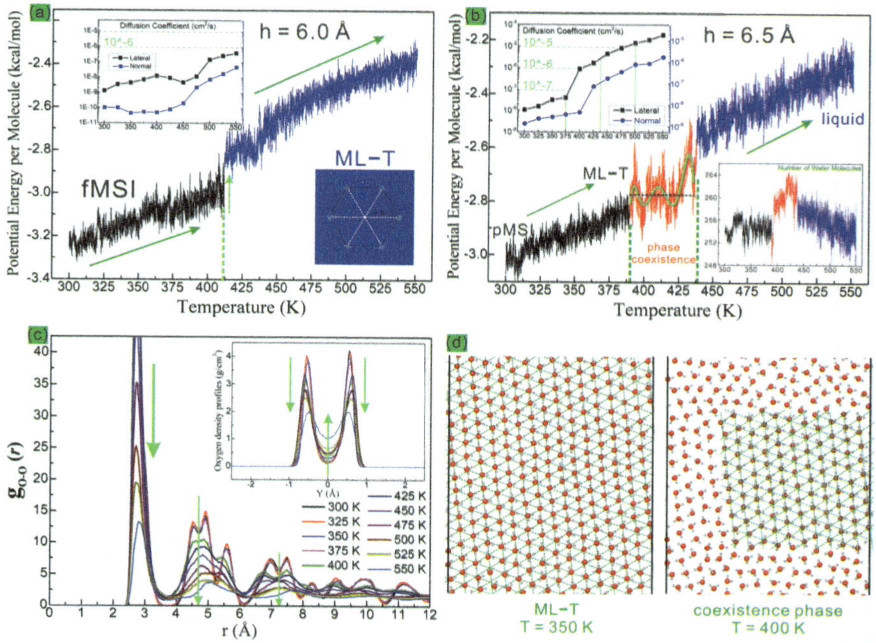

Fig. 4.5 Superheating of fMSI and pMSI under ultrahigh lateral pressure. **a** Potential energy (per molecule) during the heating process of fMSI ($h = 6.0$ Å and $P_{zz} = 3.5$ GPa). **b** Potential energy (per molecule) during the heating process of pMSI ($h = 6.5$ Å and $P_{zz} = 4.0$ GPa). **c** Oxygen-oxygen radial distribution functions. **d** ML-T and coexistence phase

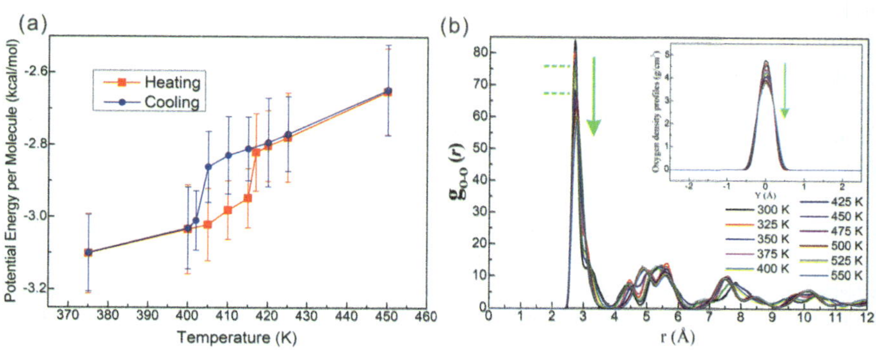

Fig. 4.6 Superheating of fMSI under ultrahigh lateral pressure. **a** Hysteresis loop in potential energy. **b** Oxygen-oxygen radial distribution functions

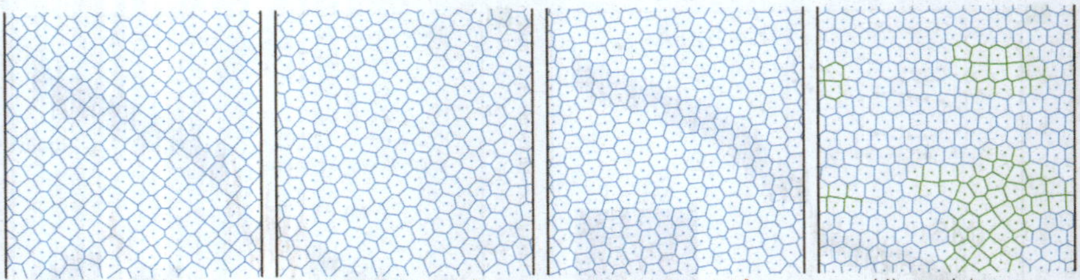

Fig. 4.7 Voronoi tessellation diagrams of several structures

ML-T. In further simulations, ML-T is hardly heated into liquid state. At 550 K, the lateral diffusion coefficient of ML-T is only ~3.8×10^{-7} cm^2/s, which is lower than that of bulk water and melted 2D water (10^{-5} cm^2/s magnitude), while higher than that of 2D ice ($<10^{-8}$ cm^2/s) [6]. The oxygen-oxygen radial distribution function in Fig. 4.6b shows that water molecules remain ordered arrangement during the superheating process. ML-T does not transform into disordered liquid even at 550 K. The values of first peaks in the $\mathbf{g}_{\text{O-O}}(r)$s show an obvious drop when the structure transforms from fMSI to ML-T, while the oxygen density profiles only display fine differences. Voronoi tessellation diagrams in Fig. 4.7a, b demonstrate the obvious structural difference between fMSI and ML-T. The polygons in Fig. 4.7b show a good hexagonal network. In the further heating process, ML-T could transform into liquid state under very high temperature (~645 K) through continuous-like phase transition. Therefore, the superheating of fMSI under ultrahigh lateral pressure is a typical discontinuous crystal-to-amorphous (solid-to-solid) scenario with a followed continuous-like solid-to-liquid transition, in which the too narrow channel and high lateral pressure are the primary factors.

In Fig. 4.5b, the superheating of pMSI (h = 6.5 Å and P_{zz} = 4.0 GPa) illustrates another extraordinary scenario. In the heating process, abrupt drop or jump is not found in the potential energy, while potential energy display an abnormal fluctuation when the temperature ranges from 390 to 400 K. We used a seven-order polynomial to fit the curve of potential energy in this temperature range, and the phase boundaries could be determined by the Maxwell construction. The fitted green curve in potential energy displays a Mayer-Wood loop [28], characterizing the coexistence phase in the superheating process. Binder et al. mentioned that the Mayer-Wood loop in energy is due to the interface free energy between coexistence phases in finite size systems [29, 30].

Meanwhile, the pressure perpendicular to graphene ($P_{\perp}$ in Fig. 4.8) also shows an abnormal nonlinear increase during this temperature range. Both of the potential energy and normal pressure manifest approximate linear increase except the phase coexistence stage, indicating the unstable feature of coexistence phase. When temperature is lower than 390 K, the number of water molecules inside the graphene channel shows a small increase and then reaches to a stable value after the formation of ML-T. Based on the change in potential energy, the transformation from pMSI

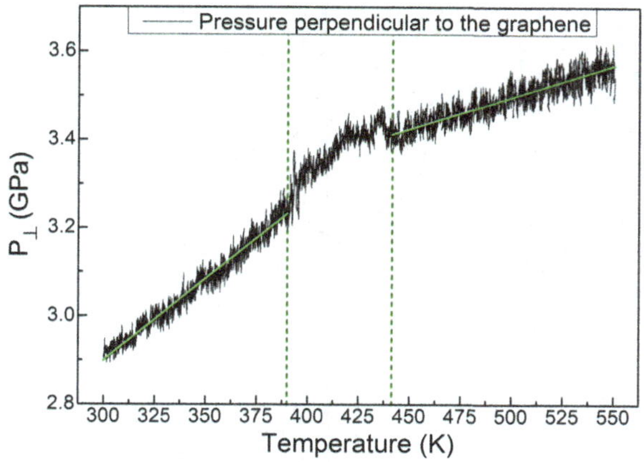

Fig. 4.8 Pressure perpendicular to the graphene during the superheating of pMSI ($h = 6.5$ Å and $P_{zz} = 4.0$ GPa)

to ML-T could be defined as a continuous-like phase transition. When the temperature is 375 K, the lateral diffusion coefficient of pMSI in graphene channel is 4.0 $\times$ 10^{-8} cm^2/s, which is lower than that of bulk water. In Fig. 4.5c, when temperature is lower than 375 K, oxygen-oxygen radial distribution functions are almost overlapped except the first sharp peaks. When temperature is higher than 440 K, the number of water molecules inside graphene channel decreases gradually with increasing temperature. The lateral diffusion coefficient is larger than 10^{-6} cm^2/s and quickly beyond 10^{-5} cm^2/s at $T = 475$ K, indicating that water molecules in the channel are in the melted liquid state. The first sharp peaks in oxygen-oxygen radial distribution functions show a rapidly decrease. Oxygen density profiles in the inset of Fig. 4.5c also indicate the liquid state of water molecules in graphene channel when temperature is beyond 440 K.

However, in the coexistence stage (390 K $\leq T \leq$ 410 K), the number of water molecules inside graphene channel reaches to the maximum value. Lateral and normal diffusion coefficients increase obviously by as much as one order of magnitude, while asynchronously with a hysteresis loop. Structural snapshots in Figs. 4.5d and 4.9 indicate that water molecules in the channel are in the coexistence state of ML-T and liquid when the temperature ranges from 390 to 410 K. Thus, all of those unusual phenomena on potential energy, diffusion coefficient, normal pressure, and the number of water molecules in the channel suggest the ML-T-liquid phase coexistence during the superheating of pMSI under ultrahigh lateral pressure. Voronoi tessellation diagram in Fig. 4.7c indicates the regular hexagonal pattern in ML-T, while some irregular pentagons and heptagons are found in Fig. 4.7d for coexistence phase. The order of the transition from ML-T to liquid with the arising of coexistence phase can be defined as a discontinuous-like transition based on the Mayer-Wood loop [28–30] and drastic oscillation in potential energy. Therefore, the melting transition of pMSI under ultrahigh lateral pressure is a continuous-to-discontinuous two-step

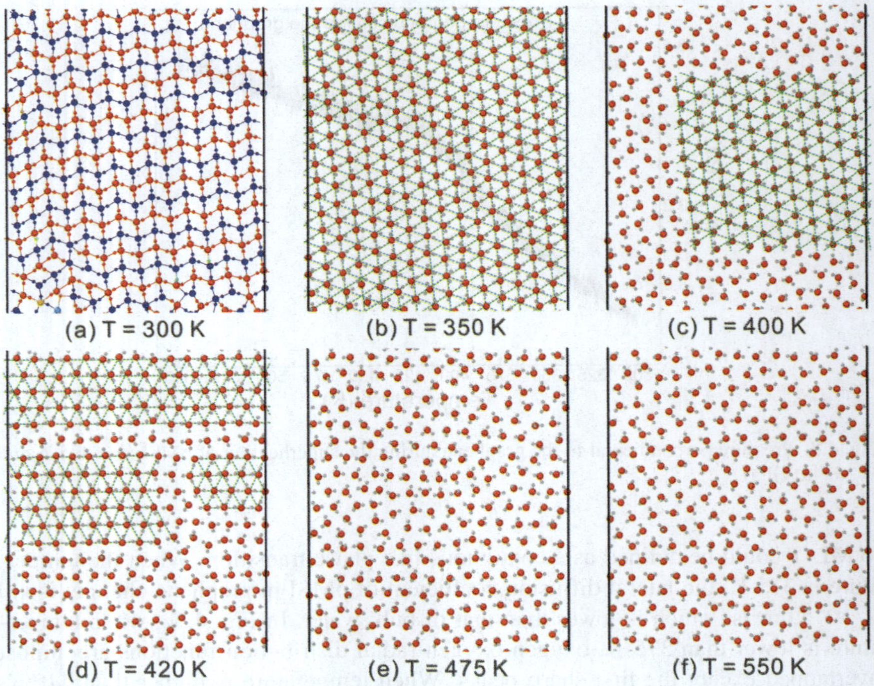

Fig. 4.9 Structural evolution during the melting of pMSI under ultrahigh lateral pressure

transition with intermediate phase coexistence. We also repeated the simulations in Fig. 4.5b to confirm the phase coexistence and abnormal fluctuation in potential energy. In most simulations, the heating rate was 5 K/ns, and we also used different heating rates (e.g., 4 and 8 K/ns) is some cases. Here, the ultrahigh lateral pressure and suitable channel width are the key factors of the arisen coexistence phase.

Note that, the melting of monolayer square-like ice under ultrahigh lateral pressure is rarely investigated in previous studies. First-order and continuous-like phases transitions were found in our MD simulations. ML-T and coexistence phase indicate that the melting of monolayer square-like ice is very sensitive to the lateral pressure. In Fig. 4.5, the main difference between the two melting scenarios is mainly due to the channel width. The too narrow nanochannel for fMSI means the highly confined effect in the thickness direction of 2D water. The coexistence phase of ML-T and liquid water should originate from that the triangular phase transformed from pMSI is not a strict monolayer structure. Oxygen density profiles in Fig. 4.5c show that ML-T transformed from pMSI also exhibits two peaks. Recently, Agrawal et al. measured the phase boundary of water confined in carbon nanotubes, revealing the ultrahigh temperature of reversible freezing/melting phase transitions in 1D confined water. The anomalous ultrahigh melting point of highly constrained monolayer ice might present some similarities with that of ice nanotube in narrow carbon nanotubes.

We also noted the Kosterlitz-Thouless-Halperin-Nelson-Young (KTHNY) theory in 2D melting [31–36]. 2D solids can melt via either a first-order phase transition or without any discontinuity in thermodynamic and may exhibit inter-mediate "x-atic" phases [36–42]. The KTHNY theory of two-step melting is based upon the behavior of topological defects in the form of dislocations and disclinations [36, 37]. Dislocations destroy translational order, which is in the hexatic phase short range (correlations decay exponential) and orientational order becomes quasi long range. Disclinations finally destroy quasi long range orientational order which becomes short range. We are unsure whether monolayer ice could be discussed in the framework of KTHNY-melting. Here, we did not extend the relevance of the melting in monolayer ice with the KTHNY theory since the complicated hydrogen bonds among water molecules and the finite simulation system, although monolayer ice also is a special 2D material. The findings in Fig. 4.5 would enhance the understanding of the superheating (melting) behavior in confined low-dimensional water/ice system and the melting in 2D materials.

4.3.4 Melting of Highly Puckered Monolayer Square-Like Ice

In our MD simulations, we also observed another superheating behavior of monolayer ice in graphene channel with a slightly large channel width. The simulation started from hpMSI (h = 7.0 Å) in 10 ns equilibrium run under the given lateral pressure (3.0 GPa) and temperature (300 K). Then the system is gradually heated to 500 K in the rate of 5 K/ns. As illustrated in Fig. 4.10, the number of water molecules inside graphene channel and the potential energy (per molecule) are used to characterize the superheating behavior of hpMSI. In the heating process, both the potential energy

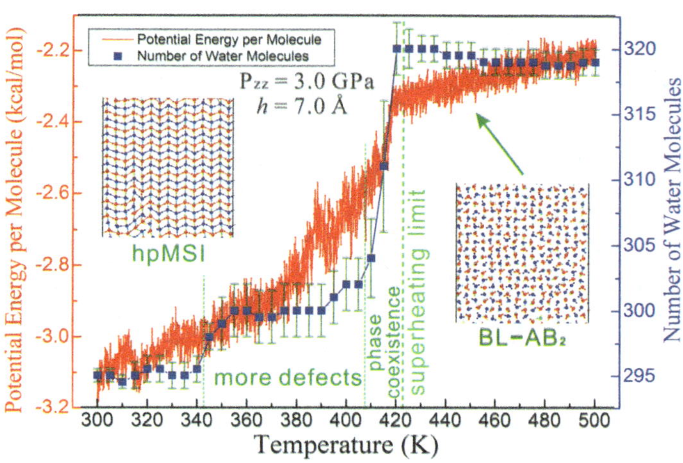

Fig. 4.10 Melting of hpMSI under ultrahigh lateral pressure

and the number of water molecules shows a sudden increase near 422 K. The sudden discontinuous changes in Fig. 4.10 indicate a first-order phase transition [4, 5] and give the superheating limit of hpMSI at $P_{zz} = 3.0$ GPa. Especially, the obvious increase in the number of water molecules indicate the formation of new phase. The insets in Fig. 4.10 show that hpMSI transforms into the AB-stacked square-like bilayer amorphous ice (BL-AB_2). BL-AB_2 can also be formed in the compression of hpMSI under ambient temperature and ultrahigh lateral pressure [14].

In the superheating of hpMSI under ultrahigh lateral pressure, we divided the heating process into four different stages based on the number of water molecules inside graphene channel. In the first stage (300–345 K), water molecules can form perfect hydrogen-bonding network except few point defects. In the second stage (345–408 K), Fig. 4.11a illustrates that the monolayer structure exhibits some defects, indicating that the ordered hydrogen-bonding network in hpMSI is gradually destroyed by the increased temperature. In this stage, the number of water molecules increases slightly, which should be attributed to that few water molecules

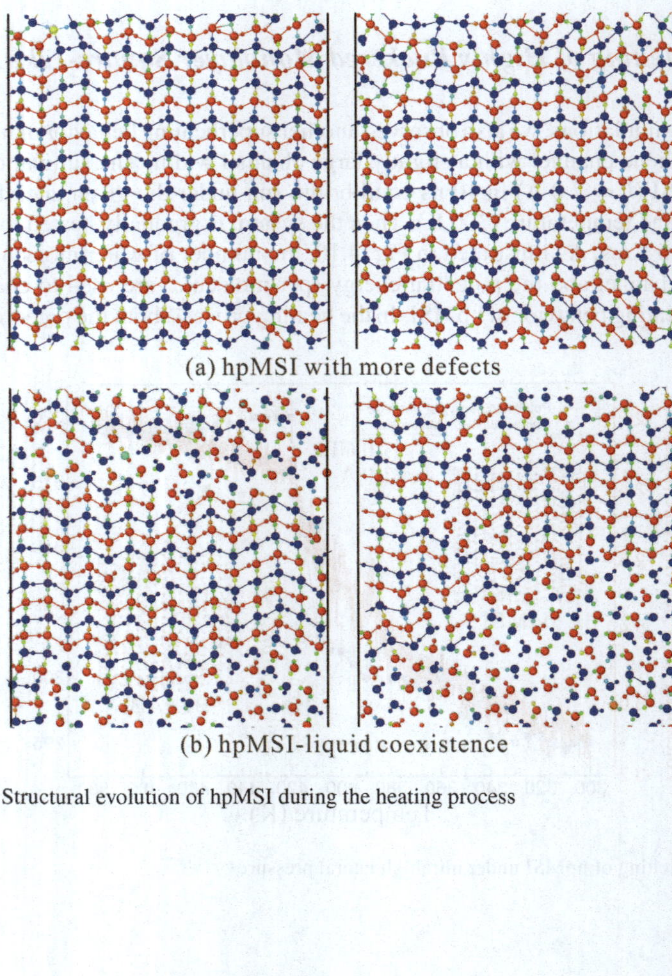

Fig. 4.11 Structural evolution of hpMSI during the heating process

enter into the destroyed region driven by high lateral pressure. In the third stage (408–422 K), both the potential energy and the number of water molecules show a sudden jump. Structural snapshots in Fig. 4.11b indicate the appearance of hpMSI-liquid coexistence in this short stage.

The coexistence structures in Fig. 4.11 are much different from the coexistence phase of ML-T and liquid in Fig. 4.9. ML-T cannot formed during the superheating of hpMSI. In Fig. 4.10, the channel width is 7.0 Å, which is slightly larger than that in Fig. 4.5b. The coexistence phase of monolayer ice and liquid water also found in MD simulations reported by Zangi and Mark [5]. The structure of hpMSI has an in-plane rhombic symmetry with respect to the positions of oxygen atoms. The hpMSI-liquid coexistence found in our simulations displays some similarity to the ice-liquid monolayer reported by Zangi and Mark using TIP5P force field [5]. In the last stage (422–500 K), the temperature is larger than the superheating limit, and the hydrogen-bonding network is destroyed entirely. Water molecules in the channel form BL-AB_2 in this stage. The ultrahigh lateral pressure and the wider channel width are the primary factors that results in this crystal-to-amorphous (solid-to-solid) phase transition.

The four melting (superheating) scenarios found in our MD simulations demonstrate that the melting phase transition of monolayer ice is very sensitive to the channel width and lateral pressure. In this current study, the effects of system temperature and lateral pressure are interactive and correlative during the superheating. The intensity (stability) of hydrogen-bonding force among water molecules is weakened by the increased temperature, and then lateral pressure and channel width determine the scenario of superheating. Under low lateral pressure, the first-order solid-to-liquid phase transition is similar with melting phase transition reported by Zangi and Mark [5]. Under ultrahigh lateral pressure, the formation of ML-T can be regarded as the compression limit of monolayer ice during the superheating. The perfect hydrogen-bonding network in monolayer square-like ice is destroyed by high temperature [4], while the ultrahigh lateral pressure prevents the arising of liquid phase. In this chapter, the formation of BL-AB_2 is the phase transition of hpMSI coupled by high temperature and ultrahigh lateral pressure during the superheating process. The hydrogen-bonding network of hpMSI is destroyed with more defects due to the increasing temperature, simultaneously, the high lateral pressure induced compression limit is the dominating factor of the formation of BL-AB_2. Correspondingly, under low lateral pressure (1.0 or 2.0 GPa), pMSI and hpMSI could directly transform into liquid phase in the heating process. The solid-to-ML-T-to-liquid scenario in 2D ice melting should be attributed to the high constraint in the systems. The underlying mechanism of different transitions in the superheating of monolayer ice is that, to a certain extent, the superheating behavior can be regarded as the temperature-pressure coupled compression limit. High lateral pressure will prevent the arising of immediate liquid, resulting in the formation of amorphous phase and phase coexistence.

4.3.5 Melting Phase Diagram of Monolayer Ice in Graphene Nanocapillaries

At last, we performed extensive MD simulations to explore the melting phase diagram of monolayer ice in graphene nanocapillaries. In MD simulations, the lateral pressure was ranging from 1.0 to 5.0 GPa, and the values of channel width were fixed at 6.0, 6.5, and 7.0 Å. Starting with the 2D ice phases from each grid point in the P_{zz}-T plane, MD simulations were carried out to measure the phase-transition temperatures in the heating processes. In the critical coexistence region, the system was relaxed in 20 ns equilibrium run and additional 5 ns for data collection. The symbols in Fig. 4.12a–c are plotted based on the simulation data, which present the temperature of phase transition under corresponding lateral pressure. Green dashes lines denote the phase boundaries, which are obtained by connect the data points. Under low lateral pressure (P_{zz} < 2.5 GPa), the phase transition is first-order solid-to-liquid melting transition. While under ultrahigh lateral pressure (P_{zz} > 2.5 GPa), phase transitions during the superheating are complex due to the ML-T and phase coexistence, corresponding to the three different scenarios illustrated in Figs. 4.5 and 4.10. Subsequently, we plotted

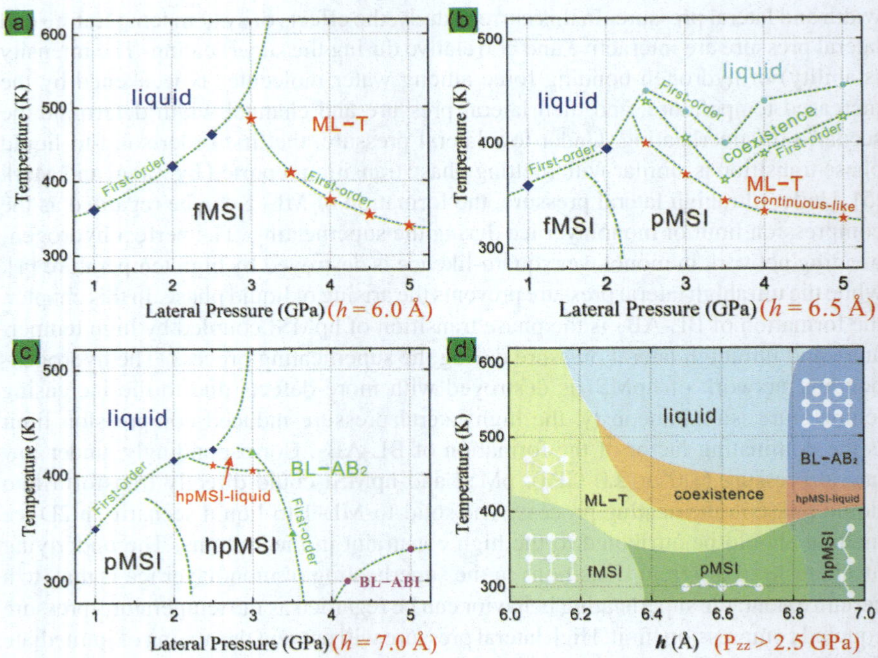

Fig. 4.12 Melting phase diagrams of monolayer ice in graphene nanocapillaries. **a–c** Phase diagrams in the P_{zz}-T plane for the cases of different channel widths. **d** Schematic phase diagram in the h-T plane for the melting of monolayer ice under ultrahigh lateral pressure

a schematic phase diagram of monolayer square-like ice under ultrahigh lateral pressure (P_{zz} > 2.5 GPa) to reveal the diverse melting transitions of monolayer ice in graphene nanocapillaries. As shown in Fig. 4.12d, the schematic phase diagram in the *h*-*T* plane is plotted based on the data in Fig. 4.12a–c.

In Fig. 4.12d, the boundaries between different colored regions were plotted based on the simulation results. We could find three different pathways of phase transition; namely, fMSI-to-ML-T, pMSI-to-ML-T-to-liquid with the ML-T-liquid coexistence, and hpMSI-to-BL-AB_2 with the hpMSI-liquid coexistence, respectively. Those phase transitions indicate that the superheating behavior of monolayer ice under ultrahigh lateral pressure is very sensitive to the channel width. In Fig. 4.12c, d, we also gave the schematic boundaries between pMSI and hpMSI. The highly puckering of monolayer square-like hydrogen-bonding network should not be neglected during the structural evolution under high lateral pressure. Although pMSI and hpMSI can be seen as the same phase, the two structures should be distinguished in phase diagram due to that the puckering degree of monolayer square-like ice is dependent on the lateral pressure.

In our recent investigations [8], we gave the evidence that lateral pressure can be as a controlling parameter to realize the first-order and continuous-like phase transitions in 2D confined water. The melting transition of monolayer ice has not been investigated in previous studies. Here, we focused on the melting transitions of monolayer ice under high lateral pressure (P_{zz} > 2.5 GPa) and proposed a schematic melting phase diagram for monolayer square-like ice.

4.4 Summary

In this chapter, we found four different scenarios of the melting of monolayer ice in graphene nanocapillaries. The superheating behavior of monolayer ice under ultrahigh lateral pressure can be regarded as the compression limit of monolayer ice during heating process. Both of first-order and continuous-like phase transitions were found in our MD simulations. Under ultrahigh lateral pressure, ML-T would be formed during the superheating of monolayer square-like ice. The anomalous solid-to-ML-T-to-liquid transition with arisen ML-T-liquid coexistence phase is found in the superheating of pMSI in narrow channel under ultrahigh lateral pressure. The observed intermediate triangular monolayer phase and coexistence phase greatly enhance the understanding of nature of melting in 2D water/ice. Whereas, under low lateral pressure, the melting of monolayer ice is a first-order phase transition. In addition, under ultrahigh lateral pressure, hpMSI could transform into square-like AB-stacked bilayer amorphous ice during the heating process. The underlying mechanism in the superheating is that the ordered hydrogen-bonding network in monolayer square-like ice is destroyed by the increasing temperature. Simultaneously, the lateral pressure determines the scenario of the superheating behavior. High lateral pressure will prevent the immediate arising of liquid phase, resulting in the formation of amorphous phase and coexistence phase. (The investigation in this chapter has been published in *J. Chem. Phys.*, 2017, **146**, 134703.)

References

1. Brovchenko I, Oleinikova A (2008) Interfacial and confined water. Elsevier, Amsterdam
2. Mochizuki K, Koga K (2015) Solid-liquid critical behavior of water in nanopores. Proc Natl Acad Sci USA 112:8221–8226
3. Zhao WH, Wang L, Bai J, Yuan LF, Yang JL, Zeng XC (2014) Highly confined water: two-dimensional ice, amorphous ice, and clathrate hydrates. Acc Chem Res 47:2505–2513
4. Koga K, Tanaka H, Zeng XC (2000) First order transition in confined water between high-density liquid and low-density amorphous phases. Nature 408:564–567
5. Zangi R, Mark AE (2003) Monolayer ice. Phys Rev Lett 91:025502
6. Bai J, Zeng XC (2012) Polymorphism and polyamorphism in bilayer water confined to slit nanopore under high pressure. Proc Natl Acad Sci USA 109:21240
7. Han S, Choi MY, Kumar P, Stanley HE (2010) Phase transitions in confined water nanofilms. Nat Phys 6:685–689
8. Zhu YB, Wang FC, Bai J, Zeng XC, Wu HA (2015) Compression limit of two-dimensional water constrained in graphene nanocapillaries. ACS Nano 9:12197–12204
9. Corsetti F, Zubeltzu J, Artacho E (2016) Enhanced configurational entropy in high-density nanoconfined bilayer ice. Phys Rev Lett 116:085901
10. Zubeltzu J, Corsetti F, Fernandez-Serra MV, Artacho E (2016) Continuous melting through a hexatic phase in confined bilayer water. Phys Rev E 93:062137
11. Algara-Siller G, Lehtinen O, Wang FC, Nair RR, Kaiser U, Wu HA, Geim AK, Grigorieva IV (2015) Square ice in graphene nanocapillaries. Nature 519:443–445
12. Zhu YB, Wang FC, Wu HA (2016) Buckling failure of square ice-nanotube arrays constrained in graphene nanocapillaries. J Chem Phys 145:054704
13. Zhu YB, Wang FC, Bai J, Zeng XC, Wu HA (2016) Formation of trilayer ices in graphene nanocapillaries under high lateral pressure. J Phys Chem C 120:8109–8115
14. Zhu YB, Wang FC, Bai J, Zeng XC, Wu HA (2016) AB-stacked square-like bilayer ice in graphene nanocapillaries. Phys Chem Chem Phys 18:22039–22046
15. Chen J, Schusteritsch G, Pickard CJ, Salzmann CG, Michaelides A (2016) Two dimensional ice from first principles: structures and phase transitions. Phys Rev Lett 116:025501
16. Chen J, Zen A, Brandenburg JG, Alfe D, Michaelides A (2016) Evidence for stable square ice from quantum Monte Carlo. Phys Rev B 94:220102
17. Corsetti F, Matthews P, Artacho E (2016) Structural and configurational properties of nanoconfined monolayer ice from first principles. Sci Rep 6:18651
18. Mario MSF, Neek-Amal M, Peeters FM (2015) AA-stacked bilayer square ice between graphene layers. Phys Rev B 92:245428
19. Neek-Amal M, Peeters FM, Grigorieva IV, Geim AK (2016) Commensurability effects in viscosity of nanoconfined water. ACS Nano 10:3685
20. Fernandez MS, Peeters FM, Neek-Amal M (2016) Electricfield-induced structural changes in water confined between two graphene layers. Phys Rev B 94:045436
21. Satarifard V, Mousaei M, Hadadi F, Dix J, Fernandez MS, Carbone P, Beheshtian J, Peeters FM, Neek-Amal M (2016) Reversible structural transition in nanoconfined ice. Phys Rev B 95:064105
22. Zhu W, Zhao WH, Wang L, Yin D, Jia M, Yang J, Zeng XC, Yuan LF (2016) Two-dimensional interlocked pentagonal bilayer ice: How do water molecules form a hydrogen bonding network? Phys Chem Chem Phys 18:14216

23. Schmidt M, Lowen H (1997) Phase diagram of hard spheres confined between two parallel plates. Phys Rev E 55:7228
24. Ayappa KG, Ghatak C (2002) The structure of frozen phases in slit nanopores: a grand canonical Monte Carlo study. J Chem Phys 117:5373
25. Koga K, Tanaka H (2005) Phase diagram of water between hydrophobic surfaces. J Chem Phys 122:104711
26. Zhao WH, Bai J, Yuan LF, Yang J, Zeng XC (2014) Ferroelectric hexagonal and rhombic monolayer ice phases. Chem Sci 5:1757–1764
27. Roman T, Groß A (2016) Polymorphism of water in two dimensions. J Phys Chem C 120:13649–13655
28. Mayer JE, Wood WW (1965) Interfacial tension effects in finite, periodic, two-dimensional systems. J Chem Phys 42:4268–4274
29. Furukawa H, Binder K (1982) Two-phase equilibria and nucleation barriers near a critical point. Phys Rev A 26:556
30. Schrader M, Virnau P, Binder K (2009) Simulation of vapor-liquid coexistence in finite volumes: a method to compute the surface free energy of droplets. Phys Rev E 79:061104
31. Strandburg KJ (1988) Two-dimensional melting. Rev Mod Phys 60:161–207
32. Kosterlize JM, Thouless DJ (1973) Ordering, metastability and phase transitions in two-dimensional systems. J Phys C: Solid State Phys 6:1181
33. Halperin BI, Nelson DR (1978) Theory of two-dimensional melting. Phys Rev Lett 41:121–124
34. Nelson DR, Halperin BI (1979) Dislocation-mediated melting in two dimensions. Phys Rev B 19:2457–2484
35. Young AP (1979) Melting and the vector Coulomb gas in two dimensions. Phys Rev B 19:1855–1866
36. Kosterlitz JM (2016) Kosterlitz-Thouless physics: a review of key issues. Rep Prog Phys 79:026001
37. Anderson JA, Antonaglia J, Millan JA, Engel M, Glotzer SC (2016) Shape and symmetry determine two-dimensional melting transitions of hard regular polygons. e-print arXiv:1606. 00687v2
38. Marcus AH, Rice SA (1996) Observations of first-order liquid-to-hexatic and hexatic-to-solid phase transitions in a confined colloid suspension. Phys Rev Lett 77:2577–2580
39. Fernandez JF, Alonso JJ, Stankiewicz J (1995) One-stage continuous melting transition in two dimensions. Phys Rev Lett 75:3477–3480
40. Bernard EP, Krauth W (2011) Two-step melting in two dimensions: first-order liquid-hexatic transition. Phys Rev Lett 107:155704
41. Kapfer SC, Krauth W (2015) Two-dimensional melting: from liquid-hexatic coexistence to continuous transitions. Phys Rev Lett 114:035702
42. Zu M, Liu J, Tong H, Xu N (2016) Density affects the nature of the hexatic-liquid transition in two-dimensional melting of soft-core systems. Phys Rev Lett 117:085702

Chapter 5
AB-Stacked and AA-Stacked Bilayer Ices in Graphene Nanocapillaries

In this chapter, we mainly focused on the bilayer phases of 2D water constrained in graphene nanocapillaries, including the AB-stacked and AA-stacked structures. In our MD simulations, we found the formations of several novel bilayer phases of confined water and gave the associated phase diagrams. Phase transitions among different bilayer ices were also discussed in detail.

5.1 Brief Introduction

The unique strength and orientation of the hydrogen bond and molecular polarity of water molecules endow a rich but still incompletely understood phase diagram of water [1–6]. The interplay between the hydrogen-bonding interaction and nanoscale confinement results in more complex phase behavior of low-dimensional water and ice [4–8]. Understanding the behavior of water/ice in nanoscale confinement is an active area of research with many technological implications. When constrained in the nanoscale environment and under ultrahigh pressure, water and ice manifest dramatic differences from their bulk counterparts in part because the van der Waals pressure can induce water-to-ice transformation [7], known as the metastability limit of the 2D liquid. The metastability limit extension of liquid water at high compression/tensile pressure or high degree of superheating/supercooling provides additional features that manifest themselves in the equilibrium states of water/ice [8–10]. The effects of confinement and associated van der Waals pressure on the mechanical stability of water/ice are yet to be comprehensively studied.

Y. Zhu, *Phase Behavior of Two-Dimensional Water Confined in Graphene Nanocapillaries*, Springer Theses,
https://doi.org/10.1007/978-981-15-7957-8_5

Polymorphic transition can occur in 2D confined water under high lateral pressure [11]. Previous studies from computational simulations had obtained various bilayer phases of 2D water confined in slit nanopores, such as bilayer hexagonal ice (BL-ice I) [11–15], bilayer amorphous ice (BL-A) [15, 16], bilayer mixed polygonal ice [10, 15, 16], bilayer Cairo pentagonal ice [10], bilayer very-high-density amorphous ice (BL-VHDA$_1$, BL-VHDA$_2$) [11, 12, 17], and square ice-nanotube arrays (BL-VHDI) [11, 12, 18]. Liquid-to-solid, solid-to-solid, and solid-to-liquid-to-solid phase transitions were found between different bilayer polymorph ice phases, including the first-order and continuous-like phase transitions. Combining DFT calculations and MD simulations, Corsetti et al. [19] gave a semi-quantitative phase diagram of stable crystalline bilayer ices in the *P-T* plane. However, to our knowledge, a comprehensive perspective of the phase diagram of bilayer water/ice in graphene nanocapillaries has not been achieved yet.

In this chapter, we mainly focused on AB-stacked and AA-stacked bilayer phases of 2D water in graphene nanocapillaries. Two square-like bilayer ice structures were found in our MD simulations; namely, AB-stacked bilayer square-like ice and AB-stacked bilayer hexagonal ice, respectively. The formations of bilayer triangular amorphous ice and bilayer square ice-nanotube ice were compared. Several phase diagrams were also obtained from our MD simulations. Phase transitions among different bilayer phases were discussed in detail.

5.2 Model and Methods

In this chapter, we mainly used the model in Fig. 2.1 to investigate the bilayer water constrained in graphene nanocapillaries. More details are presented in Sect. 2.2. In the MD simulations of AB-stacked bilayer square-like ice, the channel width was varied from 6.5 to 7.5 Å, allowing one or two monolayers of water accommodated within the graphene channel. In the MD simulations of bilayer triangular amorphous ice, the channel width was fixed at 9.0 Å. In the MD simulations of square-like bilayer hexagonal ice, smooth hydrophobic surface with the LJ 9-3 potential ($\sigma_{\text{O-wall}} = 2.5$ Å, $\varepsilon_{\text{O-wall}} = 1.25$ kJ/mol) [17, 20] was used as virtual walls.

5.3 Results and Discussion

5.3.1 AB-Stacking Bilayer Square-Like Ice in Graphene Nanocapillaries

In monolayer square-like ice, the flat hydrogen-bonding network will become puckered when the channel width is slightly enlarged. Under high lateral pressure, highly

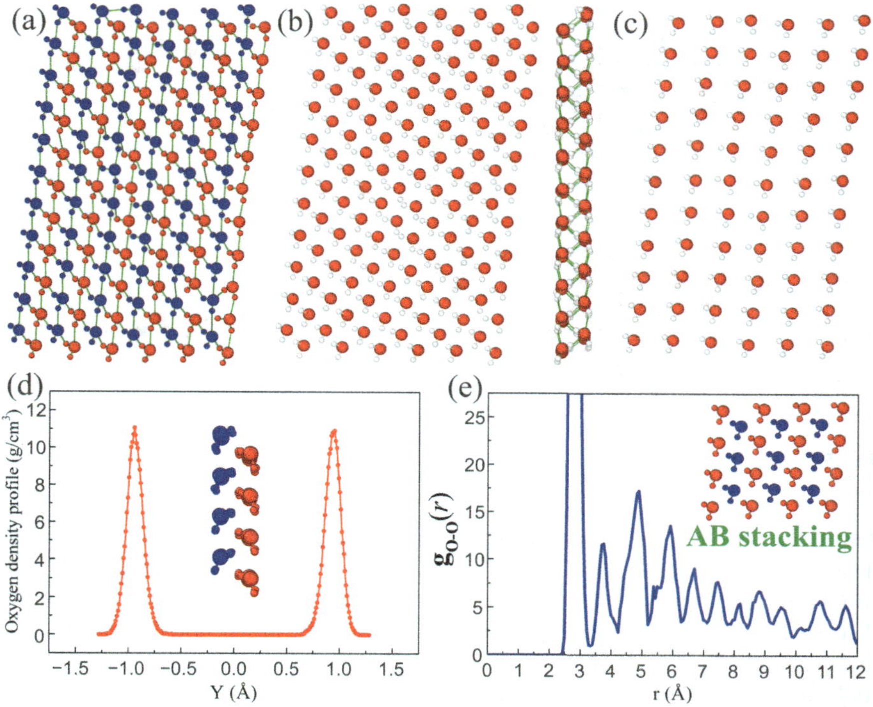

Fig. 5.1 Structural characteristics of AB-stacked bilayer square-like ice (BL-ABI). **a–c** Hydrogen-bonding network and structural snapshots. **d** Density profile of oxygen atoms in the direction perpendicular to the graphene wall. **e** Oxygen-oxygen radial distribution function

puckered monolayer square-like ice (hpMSI) could be formed in graphene nanocapillaries. Here, in Fig. 5.1, a novel bilayer ice named as BL-ABI was obtained from our MD simulations. This bilayer ice exhibits AB-stacked square-like structure, which could be transformed from puckered monolayer square-like ice (pMSI) under ultrahigh lateral pressure ($P_{zz} > 4.5$ GPa). Figure 5.1a shows that red and blue colored water molecules are in two different layers, in which water molecules in the second layer (red color) are located at the center sites among the lattice sites in the first layer (blue color). Green lines in Fig. 5.1a indicate that regular rhombic hydrogen-bonding network can be formed in BL-ABI. BL-ABI satisfies the ice rule that each water molecule could form four hydrogen bonds with its four neighbor water molecules [20]. This rhombic hydrogen-bonding network is a bilayer network, which is different from the zigzag-like puckered hydrogen-bonding network in pMSI. The side view in Fig. 5.1b shows that hydrogen bonds in BL-ABI could be classified into two types, that is, in-plane and interlayer, respectively. Note that, the interlayer hydrogen bonds are in ordered zigzag-like pattern, indicating that some hydrogen bonds are nearly parallel to the graphene surface while others are tilted with respect to the graphene wall. Interestingly, protons in BL-ABI have a regular arrangement. Each layer of

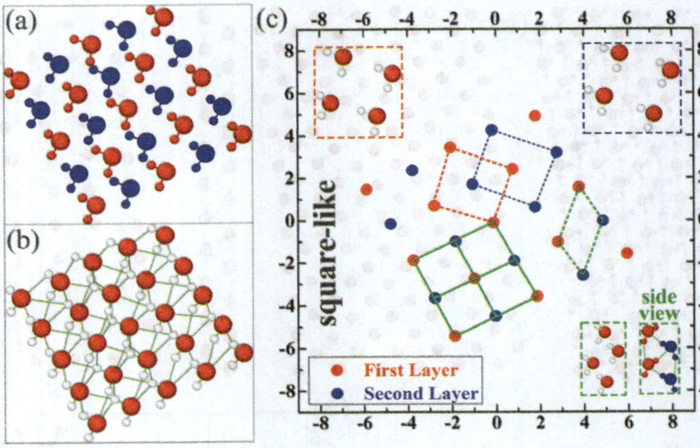

Fig. 5.2 Square-like structural features of BL-ABI

BL-ABI in Fig. 5.1c resembles the monolayer ice reported by Zangi and Mark [21]. The formation of BL-ABI is mainly due to the relatively small channel width and ultrahigh lateral pressure, resulting in higher density (Fig. 5.1d) with respect to the AA-stacked structure. The oxygen-oxygen radial distribution function in Fig. 5.1e shows that the first sharp peak is located at 2.79 ± 0.06 Å. The well-separated peaks and valleys indicate that the BL-ABI is a crystalline phase.

This AB-stacked bilayer ice exhibits the square-like structural feature although water molecules in each layer display a near-rhombic-network arrangement. From the top views in Figs. 5.1a and 5.2a, every four neighboring water molecules (two red and two blue molecules) in BL-ABI form a square-like pattern. To shed light on the structural feature of BL-ABI, we randomly selected a small part of BL-ABI in Fig. 5.2 as an example. The center of this section is relocated as the origin of coordinates in Fig. 5.2c, then the positions of oxygen atoms are plotted on this coordinate plane. The red and blue points in Fig. 5.2c denote two different layers in BL-ABI. Blue and red dashed boxes display the square-like (slightly rhomboidal) pattern in each layer. The square network formed by green solid lines indicates the square configuration of every four neighboring water molecules in BL-ABI. The rhombus formed by green dashed lines denote the lateral rhombic structure between two layers. The side view in the inset of Fig. 5.2c indicates that water molecules in two layers can form interlayer zigzag-like and in-plane hydrogen bonds. Based on the coordinates of oxygen atoms, the length of red and blue dashed lines is ~2.74 Å, and the length of green solid lines is 2.0 ± 0.08 Å. In the green-dotted rhombic unit (considering the interlayer distance), the lengths of two sides are ~2.735 Å and ~2.84 Å, respectively. The interlayer distance of the two layers is ~1.96 Å.

In the following, we investigated the phase transition between pMSI and BL-ABI in detail. We carried out independent MD simulations of compression and decompression processes (h = 7.0 Å, T = 300 K, and 1.0 GPa < P_{zz} < 6.0 GPa). The change of lateral pressure was in a step of 0.1 GP/ns. As shown in Fig. 5.3,

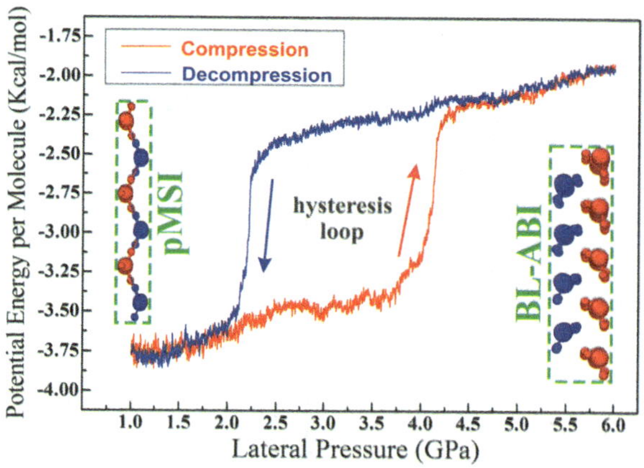

Fig. 5.3 Potential energy (per molecule) and its hysteresis loop during the compression and decompression processes. The insets show the lateral configurations of pMSI and BL-ABI, respectively

red and blue lines give the potential energy (per molecule) during the compression and decompression processes, respectively. In the compression simulation, potential energy increases gradually and then jumps rapidly at ~4.1 GPa. This sudden increase in potential energy is mainly due to the transformation from pMSI to BL-ABI. Before the formation of BL-ABI, monolayer liquid water first forms pMSI when the lateral pressure is beyond the compression-limit (P_{zz} > 0.6 GPa) [22], which then forms hpMSI with increasing lateral pressure (P_{zz} < 3.8 GPa). In the decompression process, potential energy decreases gradually and then drops rapidly at ~2.2 GPa, overlapping with the path in compression process. Note that, between 2.2 GPa and 4.1 GPa, the number of layers in 2D ice during the compression and decompression processes is changed, giving the compression/tension limit of the 2D monolayer/bilayer ice. In addition, during the transformation from pMSI to BL-ABI, the arrangement of hydrogen atoms is changed. In pMSI, the distribution of protons is along the rows and columns, while in BL-ABI it shows an ordered arrangement.

To further understand the phase transition between pMSI and BL-ABI, we captured the structural snapshots of water molecules in the channel under different lateral pressure. Each simulation was controlled in $NP_{zz}T$ ensemble for 25 ns under the given lateral pressure. Figure 5.4 illustrates six different structures from pMSI to BL-ABI. Note that, the AB-stacked bilayer amorphous (BL-AB_2) was found between hpMSI and BL-ABI. Although BL-AB_2 exhibits the similar square-like lattice, the configuration of hydrogen atoms is disordered. Under low lateral pressure (~1.0 GPa), pMSI exhibits perfect square-like hydrogen-bonding network. With increasing lateral pressure, pMSI becomes to be more puckered, and then hpMSI shows more clearly rectangular and rhombic characteristics. At last, the AB-stacked bilayer crystalline

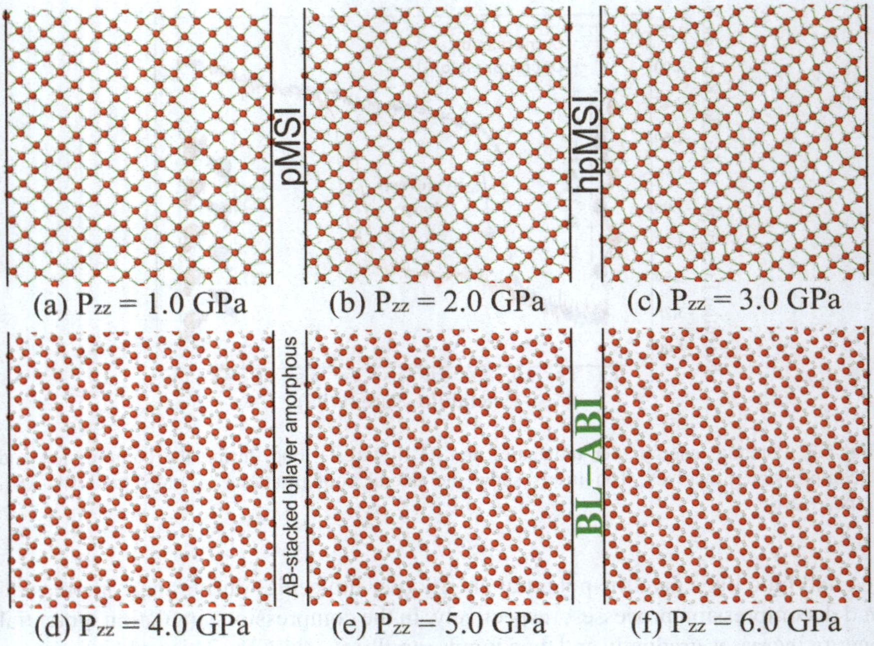

Fig. 5.4 Structural snapshots during the phase transition from pMSI to BL-ABI

ice displays a compact square-like pattern where the hydrogen atoms exhibit a high concordance. The stacking order of oxygen atoms in BL-AB_2 and BL-ABI is similar with that in bilayer structures predicted hard or soft balls between two parallel plates [23, 24].

Based on the structural evolution in Fig. 5.4, the transformation from pMSI to BL-ABI is a crystal-to-amorphous-to-crystal phase transition, which is similar to the phase transition path from bilayer triangular ice to ABA-stacked trilayer square-like ice [25]. To illustrate this crystal-to-amorphous-to-crystal phase transition, we computed the oxygen density profile, mean-squared displacement (MSD), and oxygen-oxygen and oxygen-hydrogen radial distribution functions at different lateral pressure. In Fig. 5.5a, the peak value of oxygen density for hpMSI is only half of that for BL-ABI, indicating that BL-ABI is a very-high-density phase. MSD results in Fig. 5.5b demonstrate that the diffusion coefficient of BL-ABI is lower than that of hpMSI (~3.0×10^{-9} cm^2/s). The oxygen-oxygen and oxygen-hydrogen radial distribution functions can also show the difference among three phases, in which the locations of first peaks are 2.79 ± 0.06 Å and 1.75 ± 0.03 Å, respectively.

Ice formation and ice structures in nanoscale confinement depend strongly on lateral pressure and pore width [22]. We performed extensive MD simulations to obtain the compression-limit phase diagram in h-P_{zz} plane with the lateral pressure ranging from 0.1 to 6.0 GPa and channel width ranging from 6.5 to 7.5 Å. The marked symbols in Fig. 5.6 denote the compression limit of the corresponding phases. The

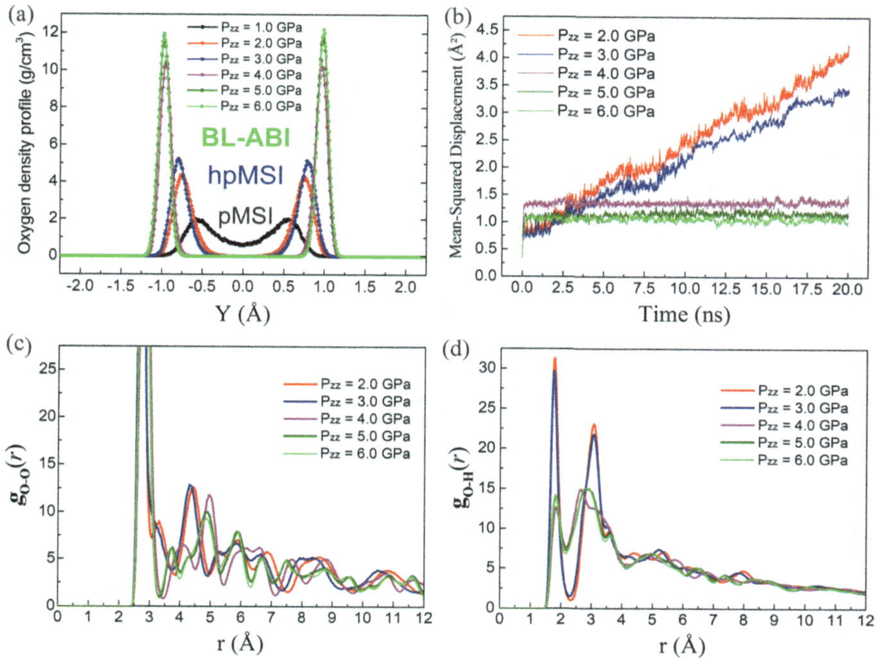

Fig. 5.5 Phase transition from pMSI to BL-ABI. **a** Oxygen density profiles. **b** Mean-squared displacement. **c** Oxygen-oxygen radial distribution function. **d** Oxygen-hydrogen radial distribution function

compression-limit curves (solid lines) are approximately drawn by connecting the compression-limit points. The dashed lines are schematic phase boundary lines. As illustrated in Fig. 5.6, 2D liquid water is transformed into pMSI when the lateral pressure is beyond the compression limit (black line with square symbols). Under much larger lateral pressure, pMSI will be transformed to BL-ABI when the channel width is smaller than 7.2 Å, or to BL-AB phase in a relatively wide channel ($h >$ 7.3 Å). We overlapped the compression-limit lines of the amorphous (blue line) and BL-AB (olive line) at the point (7.2 Å, 3.2 GPa). But the blue line is due to a vertical path (hpMSI-to-amorphous), and the olive line is due to a horizontal path (increasing the pore width) [22]. The two paths are different. The region between the blue and red lines denotes the amorphous phase during the crystal-to-amorphous-to-crystal phase transition. Oxygen atoms in BL-ABI and the amorphous phase exhibit square configurations, while those in BL-AB show rhombic arrangements.

We also investigated the superheating behavior of BL-ABI in the graphene nanocapillaries ($h = 7.0$ Å). The initial state in MD simulations was BL-ABI under given lateral pressure (5.0 GPa) and temperature (300 K). Then the system was heated to the given temperature within $NP_{zz}T$ ensemble for 20 ns. The insets in Fig. 5.7a shows that BL-ABI is transformed to BL-AB_2 during the superheating

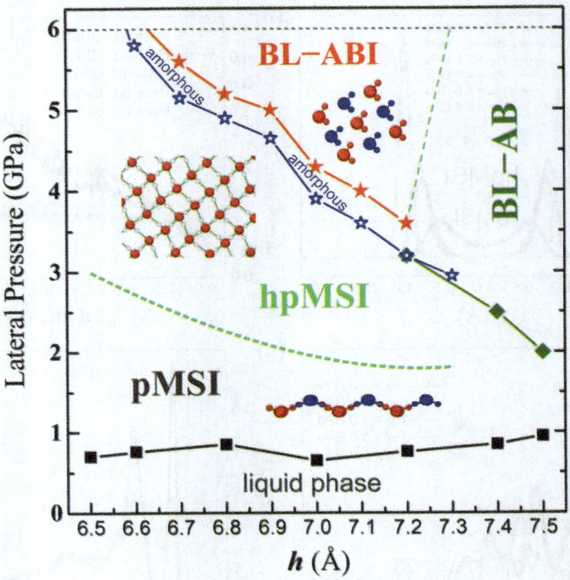

Fig. 5.6 Compression-limit phase diagram of pMSI under ambient temperature

process, and potential energy (per molecule) is not fully linear with increasing temperature. Oxygen atoms still maintain the square-like pattern, while the configuration of hydrogen atoms is disordered. Figure 5.7b shows that the values of the first sharp peak in oxygen-hydrogen radial distribution functions decrease gradually with increasing temperature, indicating that the configuration of hydrogen atoms is not as regular as that in BL-ABI. Oxygen and hydrogen density profiles are plotted in Fig. 5.7c, d, in which the peak values decrease gradually with increasing temperature. Therefore, the superheated phase of BL-ABI is the proton-disordered structure.

5.3.2 AA-Stacking Bilayer Ice in Graphene Nanocapillaries

In the phase transition of highly confined water, temperature and pressure are two obviously different but equivalently important factors. At ambient temperature, our MD simulations demonstrated that bilayer water can transform into BL-AB and BL-AAI when the lateral pressure is beyond corresponding compression limit. In previous studies, another bilayer crystalline ice with regular square ice nanotube array (BL-VHDI) was reported. To gain more insight into bilayer water, we here considered the phase transition at different temperatures and the channel width was fixed at $h = 9.0$ Å. Figure 5.8a shows the simulation results of bilayer water at $T =$ 240 K during the compression process (0.1 GPa $\leq P_{zz} \leq$ 5.0 GPa), which are much

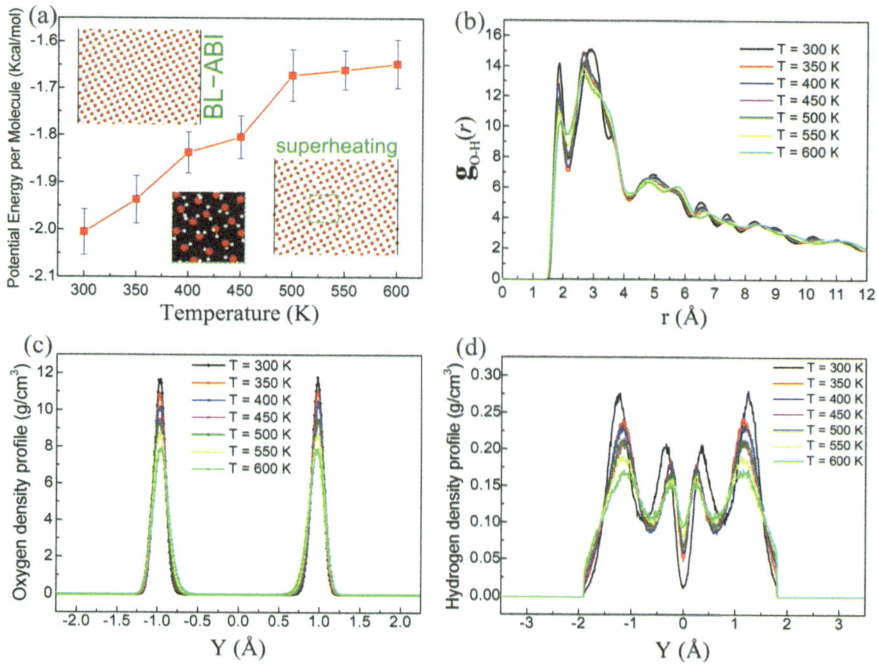

Fig. 5.7 Superheating of BL-ABI. **a** Potential energy as a function of temperature. **b** Oxygen-hydrogen radial distribution functions. **c** Oxygen density profiles. **d** Hydrogen density profiles

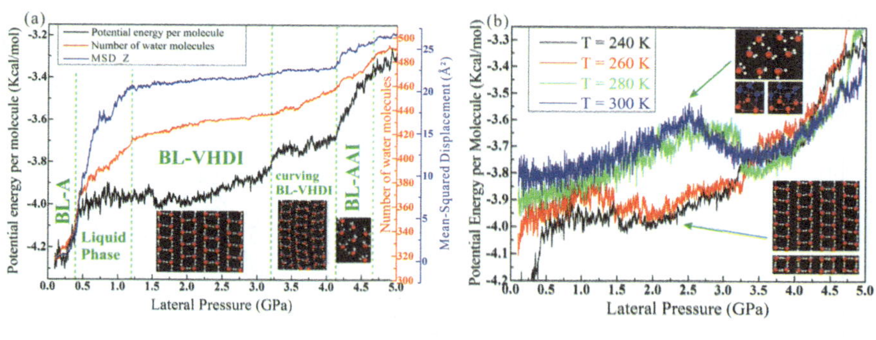

Fig. 5.8 Phase transitions of bilayer water in graphene nanocapillaries at different temperatures. **a** Compression process of bilayer water at $T = 240$ K and $h = 9.0$ Å. **b** Potential energy curves at different temperatures

different from the phase transition of bilayer water at ambient temperature. Five sub-regions can be classified in Fig. 5.8a based on the structures; that is, BL-A, bilayer liquid, BL-VHDI, curving BL-VHDI, BL-AAI, respectively. During the compression process, BL-A is firstly formed under low lateral pressure ($P_{zz} < 0.35$ GPa). With increasing lateral pressure, the potential energy and MSD curves are increased at a faster rate until the BL-VHDI is formed ($P_{zz} > 1.2$ GPa). The transformation from BL-A to BL-VHDI is a typical solid-to-liquid-to-solid phase transition, in which an intermediate liquid state arises between two solid phases. In this Oswald stage [11, 16], the diffusion coefficient of water molecules inside the graphene channel is in the magnitude of 10^{-7} cm/s^2. In the further compression process, ice nanotubes in BL-VHDI are bended ($P_{zz} > 3.2$ GPa) and then BL-VHDI transforms into BL-AAI ($P_{zz} > 4.15$ GPa). When the lateral pressure is beyond 4.70 GPa, another solid-to-liquid-to-liquid phase transition between BL-AAI and TL-ABAI would be observed in MD simulations [25].

To compare the dependence of bilayer water on temperature, we summarized potential energy curves at 240 K, 260 K, 280 K, and 300 K in Fig. 5.8b. The changes in potential energy curves for low and high temperatures are obviously different. When the lateral pressure is between 1.50 and 3.20 GPa, large jump is found between the potential energy curves at 260 K (red) and 280 K (green), indicating that the phase transition of bilayer water is dependent on the temperature. The distinct difference among potential energy curves also indicates that BL-VHDI is in a low-energy state related to BL-AAI. For low temperatures ($T = 240$, 260 K), potential energy decreases between 1.20 and 2.20 GPa, and then potential energy increases gradually during the further compression process. Whereas, for high temperature ($T = 280$, 300 K), potential energy shows a drop between 2.50 and 3.50 GPa, and then the curves are overlapped in the further compression process. The transformation from bilayer liquid to BL-VHDI is a first-order phase transition [11], while that from bilayer water to BL-AAI is a continuous-like phase transition [17, 22]. The large difference in potential energy curves in Fig. 5.8b indicates that a temperature threshold should exist in the range of 260–280 K. Then, we carried out MD simulations of bilayer water at other temperatures. In Fig. 5.9, potential energy curve at 275 K is different

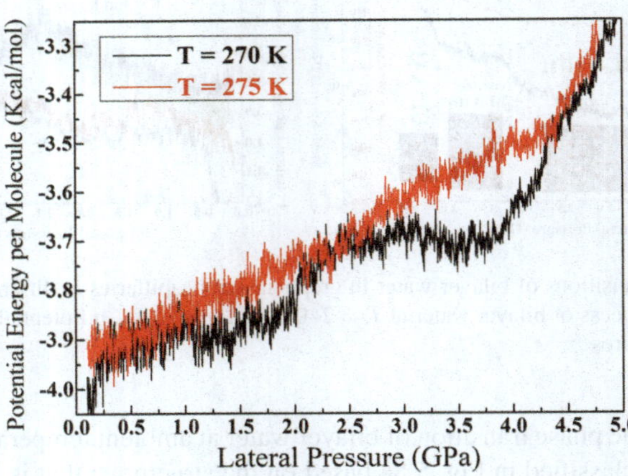

Fig. 5.9 Potential energy (per molecule) of bilayer water at 270 and 275 K during the compression process

from both of that at 260 and 280 K. Therefore, the temperature of 275 K might be the threshold between two different types of compression-limit phase transitions of bilayer water.

In our MD simulations, we found that the phase transition from BL-A to BL-VHDI and then to BL-AAI is always existed when the temperature is lower than 275 K. While when temperature is beyond 275 K, the phase transition is a direct transformation from bilayer water to BL-AAI. Figure 5.10 illustrates two different paths of bilayer water during compression phase transition. The red and blue arrows denote high and low temperatures, respectively. To gain more insight into the compression limit of bilayer water, we carried out extensive MD simulations to explore the phase diagram of bilayer water in the T-P_{zz} plane, in which the temperature range was 100–400 K, and the range of lateral pressure was 0.1–5.0 GPa. The channel width was fixed at 9.0 Å. The compression limit of bilayer water at each point of temperature was determined by the observation of phase transition in MD simulations. The lateral pressure in the compression process was increased in a step of 0.2 GPa/ns. In each compression simulation, the simulation time was 30 ns, including 5 ns initial equilibrium run.

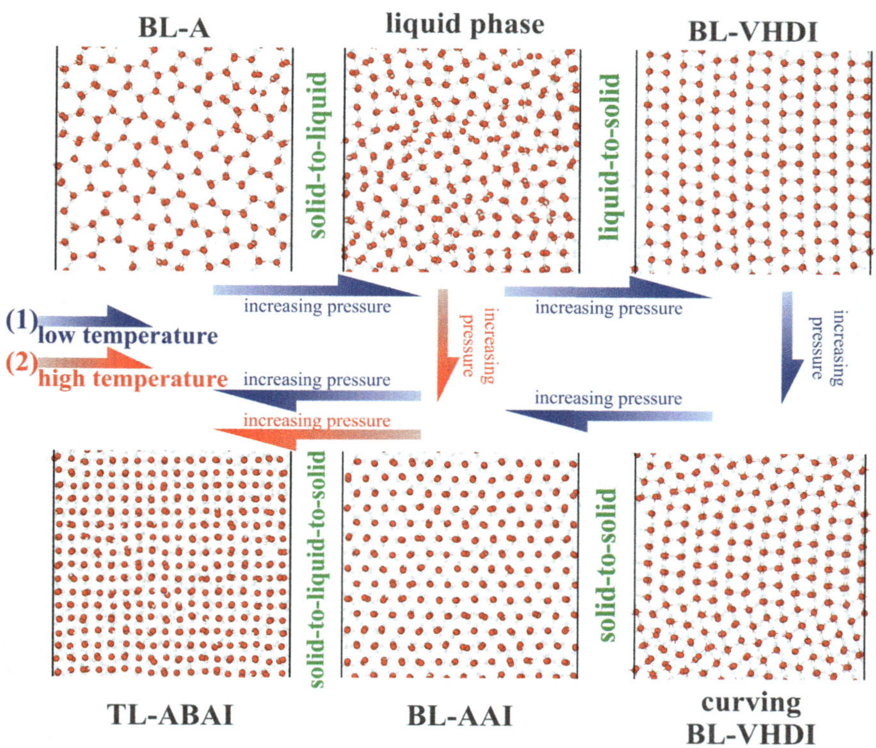

Fig. 5.10 Structural evolutions during the compression phase transition of bilayer water under low and high temperature

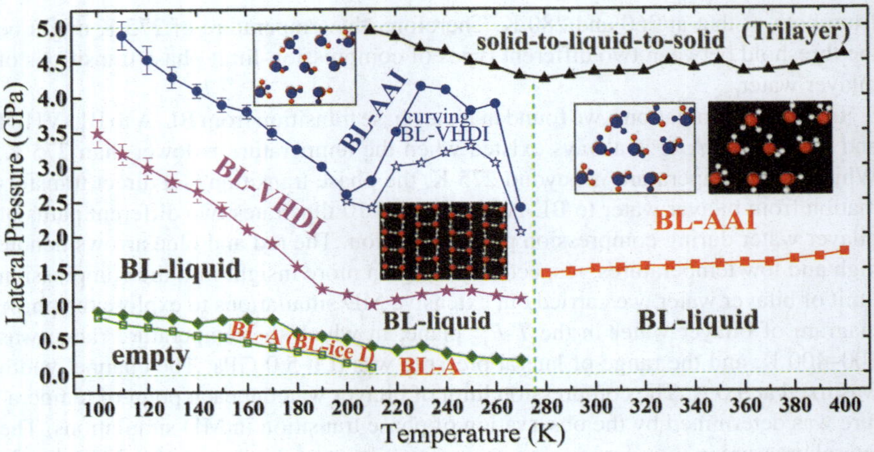

Fig. 5.11 Compression-limit phase diagram of bilayer water in graphene nanocapillaries ($h = 9.0$ Å)

The compression-limit phase diagram of bilayer water is plotted in Fig. 5.11, in which different phases are separated by the compression-limit curves and marked symbols denote the compression-limit values at corresponding temperature. This phase diagram can clearly reveal the phase transitions of bilayer water during compression process. There are three different AA-stacked phases, namely, BL-A, BL-VHDI, and BL-AAI. BL-A could transform into BL-ice I through a long-time annealing [11]. BL-A and BL-ice I could exist at low temperature and low lateral pressure. When temperature is lower than 275 K, water molecules inside graphene channel would firstly form BL-A and then transform in BL-VHDI with an intermediate bilayer liquid phase. With increasing lateral pressure, BL-VHDI would bend first and then transform into BL-AAI. When the temperature is lower than 215 K, water molecules cannot enter into the channel at lower lateral pressure. It indicates that the low temperature and lateral pressure cannot provide the needed driving force to overcome the energy barriers at the entrances, resulting in that the water molecules in reservoirs cannot be pushed into the graphene channel. When temperature is larger than 275 K, BL-A cannot be observed due to that the temperature is higher than the melting point of BL-A [11]. Water molecules inside graphene channel form bilayer liquid and then turn into BL-AAI at high lateral pressure. Under ultrahigh lateral pressure, TL-ABAI will be formed through solid-to-liquid-to-solid phase transition [25]. In Fig. 5.11, green dashed line marked boundary indicates the temperature threshold (~275 K) between BL-VHDI and BL-AAI. This phase diagram also directly reflects the structural evolution and phase transition path of AA-stacked bilayer phases in Fig. 5.10.

We also investigated the superheating of BL-VHDI. Figure 5.12 shows the superheating process of BL-VHDI from 260 to 360 K within a 20 ns MD simulation. With the increase of temperature, BL-VHDI turns into curving BL-VHDI and then

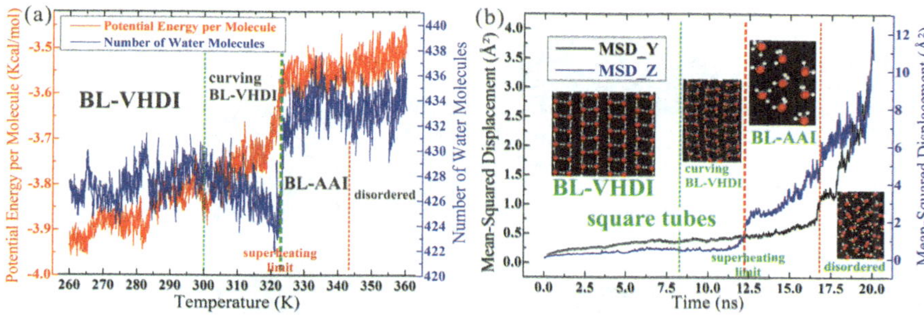

Fig. 5.12 Superheating behavior of BL-VHDI ($P_{zz} = 2.0$ GPa and $h = 9.0$ Å)

transforms into BL-AAI when the temperature is beyond the superheating limit. The transformation from BL-VHDI to BL-AAI is a solid-to-solid phase transition. Interestingly, the structural evolution of BL-VHDI in the superheating process is similar with that in the compression process. Ice nanotubes in BL-VHDI are bended first before the formation of BL-AAI. Under high temperature (>343 K), BL-AAI transforms into a disordered phase and water molecules are not in AA stacking order. In Fig. 5.12a, potential energy (per molecule) shows a sudden jump at the superheating limit (~323 K), indicating that the ordered hydrogen-bonding network is destroyed by high temperature. The MSD_Z curve in Fig. 5.12b also shows a jump at the superheating limit, while the MSD_Y curve maintains the same slope during the phase transition from BL-VHDI to BL-AAI. In addition, we also simulated the melting of BL-AAI at $T = 300$ K. Potential energy increases gradually during the superheating of BL-AAI. Thus, phase transitions in the superheating of BL-VHDI and BL-AAI are much different. The transformation from BL-VHDI to BL-AAI during the superheating process is a first-order phase transition, while the superheating of BL-AAI is a continuous-like phase transition.

During the superheating of BL-VHDI, the number of water molecules inside the graphene channel shows an abnormal change near the superheating-limit temperature. As shown in Fig. 5.12a, the number of water molecules inside the channel decreases gradually during the bending of ice nanotubes. When the temperature is close to the superheating limit, the number of water molecules shows a rapid jump and curving BL-VHDI transforms into BL-AAI. In this superheating process, the high temperature would induce the thermal expansion of ice-nanotube array. Adjacent ice nanotubes would stagger much more and high lateral pressure results in the bending of ice nanotubes. Subsequently, when the temperature is close to the superheating limit of BL-VHDI, ice nanotubes are unstable and then damaged. Therefore, temperature and lateral pressure is interactive and correlative in the superheating process of BL-VHDI.

Another series of MD simulations were carried out to obtain the superheating-limit diagram of BL-VHDI in the P_{zz}-T plane. Each simulation was started from the BL-VHDI phase at a fixed lateral pressure, and the temperature was increased from 260 K in a step of 5 K/ns. Potential energy and MSD were used to determine

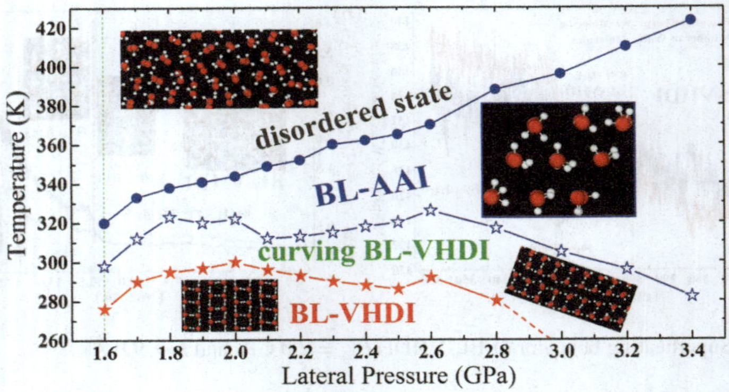

Fig. 5.13 Superheating-limit phase diagram of BL-VHDI in graphene nanocapillaries ($h = 9.0$ Å)

the temperature of phase transition. Figure 5.13 illustrates the superheating-limit diagram of BL-VHDI in graphene nanocapillaries. The superheating-limit curve of BL-VHDI exhibits local minimum, while that of BL-AAI shows a tendency of monotonous increase. Here, in the phase transitions of bilayer water/ice ($h = 9.0$ Å), temperature and lateral pressure are interactive and correlative. The compression-limit and superheating-limit (phase) diagrams can directly demonstrate the structural transformation among different AA-stacked bilayer ice and provide some intriguing phenomena between the bilayer tube phase and triangular phase. Moreover, the phase transitions shown in Fig. 5.12 indicate that the melting of BL-VHDI via a two-stage process: first a first-order transition to the BL-AAI, followed by a continuous-like transition to the disordered state.

Note that the structural transformation from BL-VHDI to BL-AAI with the intermediate bended ice nanotubes was observed both in compression and superheating processes. During the phase transition from BL-VHDI to BL-AAI, as shown in Fig. 5.14, the square ice nanotubes are bended gradually with the increase of lateral pressure or temperature. The bended ice nanotubes are more like the buckling state of compressive bar in mechanics. Here, we proposed a novel transformation mechanism between the two AA-stacked bilayer ice polymorphs.

The stability of compressive bar is an important problem in the mechanics of materials and structures. Buckling failure is a common failure mode in beam column and shell structures in engineering. In our MD simulations, BL-VHDI can be regarded as ice-nanotube arrays under axial loading (lateral pressure). The structure of curving BL-VHDI is widely existed and cannot be ignored in the compression and superheating of BL-VHDI. Therefore, in this structural evolution, each ice nanotube in BL-VHDI can be seen as a compressive bar. Lateral pressure is a primary factor in the phase transition and induces the bending of ice nanotubes. We also calculated the oxygen-oxygen radial distribution functions of BL-VHDI and BL-AAI. As shown

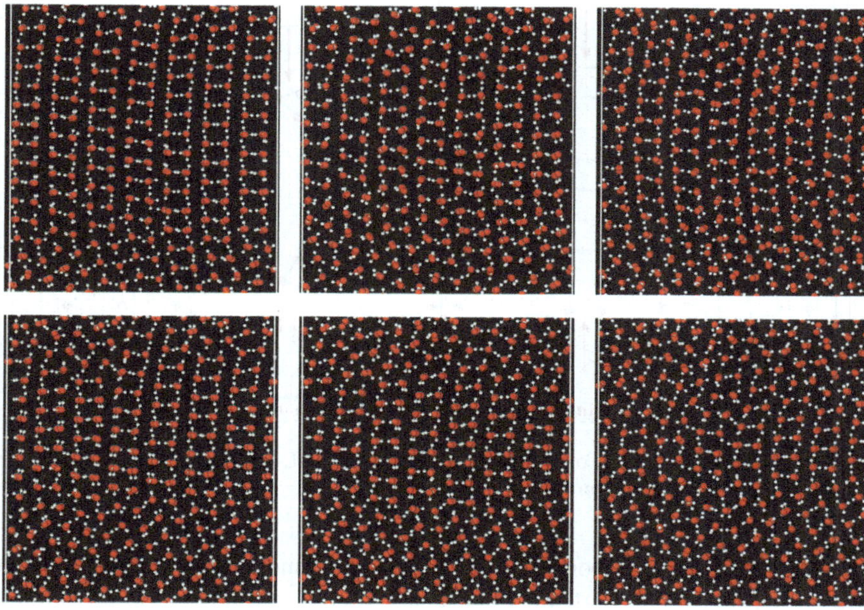

Fig. 5.14 Structural evolution of bended ice nanotubes during the transformation from BL-VHDI to BL-AAI

in Fig. 5.15, the first sharp peaks of both two phases are located at 2.79 ± 0.06 Å. Oxygen atoms in this structural evolution always maintain the AA stacking order. Thus, we could use a planar schematic diagram in Fig. 5.16 to represent the two ice structures.

In the compression process of BL-VHDI ($T < 275$ K), the gradually increased lateral pressure equivalently enlarges the "axial loading" on ice nanotubes and then directly results in the bending of straight ice-nanotube columns. When the lateral

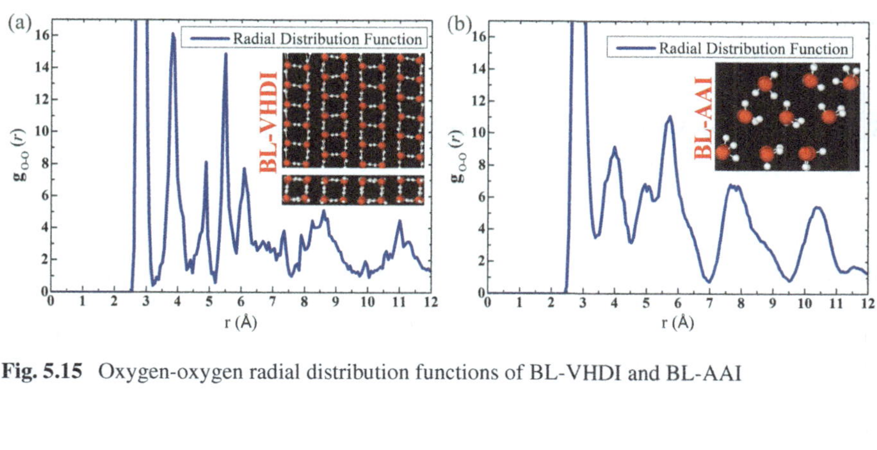

Fig. 5.15 Oxygen-oxygen radial distribution functions of BL-VHDI and BL-AAI

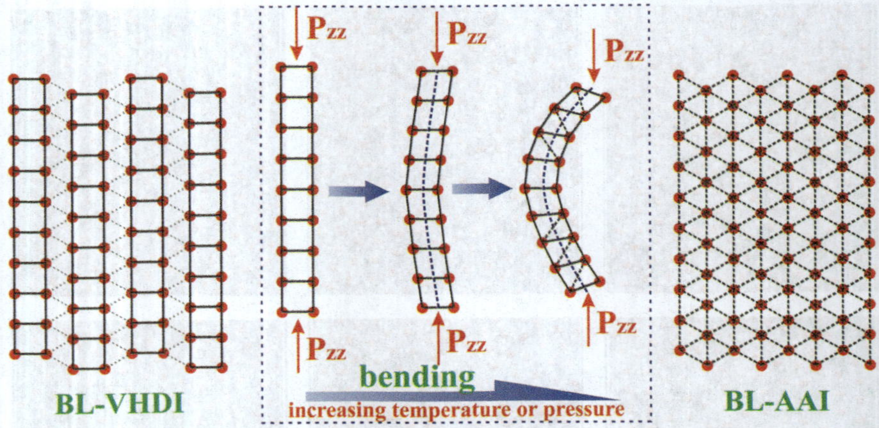

Fig. 5.16 Schematic diagram of structural evolution from BL-VHDI to BL-AAI during the compression and superheating processes

pressure is beyond corresponding compression limit, most of the square elements in ice-nanotube columns turn into rhombs or triangles, and simultaneously, curving BL-VHDI transforms into BL-AAI. The critical pressure of this structural transformation in compression is the compression-limit boundary between BL-VHDI and BL-AAI (Fig. 5.11). In the superheating process, the gradually increased temperature reduces the stability of ice nanotubes. In other words, the temperature determines the intensity of hydrogen-bonding force among water molecules in ice nanotubes. With increasing temperature, the fixed lateral pressure will be beyond corresponding compression limit of BL-VHDI at high temperature. When the temperature is beyond the superheating limit, the curving BL-VHDI transforms into BL-AAI accompanied with most of square elements turning into rhombs and triangles. The critical temperature in superheating is the superheating-limit boundary between bended BL-VHDI and BL-AAI (Fig. 5.13). We also used large model to verify this structural evolution from BL-VHDI to BL-AAI. The critical pressure of the buckling of ice nanotubes should be dependent on the size of model.

5.3.3 Bilayer Square-Like Ice with AB-Stacked Elongated Hexagonal Rings

In our MD simulations, we also found another bilayer square-like ice at low temperature. To better display the structure and phase transition of this bilayer ice, we and co-workers used virtual parallel walls with the LJ 9-3 potential [20], which is different from the main model in Sect. 2.2. The realistic structure of this bilayer square-like ice consists of two hexagonal monolayers with the AB-stacking order (named as BHI-AB), where the hexagonal rings are slightly elongated with two unequal inner angles,

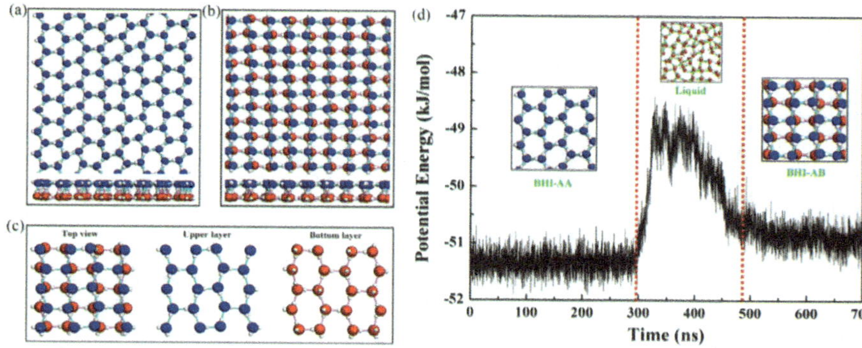

Fig. 5.17 Structures and phase transition of two bilayer ice phases. **a** Bilayer hexagonal ice with the AA-stacking order (BHI-AA). **b** Bilayer square-like ice with the AB-stacking order of elongated hexagonal rings (BHI-AB). **c** Structural features of BHI-AB. **d** Phase transition from BHI-AA to BHI-AB via slowly reducing the width of nanoslit

107° and 146° (rather than 120°). Similar structure was previously reported by Jia et al. [26] while without detailed investigation on its phase transition. Herein we and co-workers mainly concerned the structural feature of BHI-AB and associated solid-to-liquid-to-solid phase transitions.

Figure 5.17 illustrates two different bilayer ice phases; that is, BHI-AA and BHI-AB, respectively. Here, the AA stacking means that the positions of hexagonal rings in the upper layer overlap with those in the lower layer [7]. While the AB stacking means that the positions of elongated hexagonal rings in the upper layer are staggered with those in the lower layer by the half length of hexagonal unit. More interestingly, BHI-AB exhibits a top-view pattern that resembles a bilayer square-like ice in the AA-stacking order. However, in Fig. 5.17c, BHI-AB actually consists of two AB-stacked hexagonal monolayers, each exhibiting elongated hexagonal rings. BHI-AB satisfies the ice rule in that every water molecules is a double donor and a double acceptor of hydrogen bonds to its four neighbor molecules [20]. The interlayer molecular interaction in BHI-AB is relatively strong because the two layers are connected via hydrogen bonds between every pair of two opposing water molecules. These interstratified hydrogen bonds are likely the primary reason for the high stability of this AB-stacked ice structure [22]. Each pair of two opposing water molecules in the two layers forms an approximate AA-stacking configuration in the top view. Therefore, a square-like pattern of the whole structure of BHI-AB can be seen in the top view after taking the time-averaging effect into account.

We also found that BHI-AB can be transformed from BHI-AA via slowly reducing the width of nanoslit. Note that this BHI-AA to BHI-AB transformation could show the Oswald staging phenomenon [11, 16], in which an intermediate liquid state arises during the solid-solid phase transition. Figure 5.17d displays this solid-to-liquid-to-solid phase transition at the given channel width of 7.7 Å, temperature of 210 K, and lateral pressure of 450 MPa. The time-dependent potential energy (per molecule) also reflects the Oswald staging behavior, where the liquid state has the highest potential

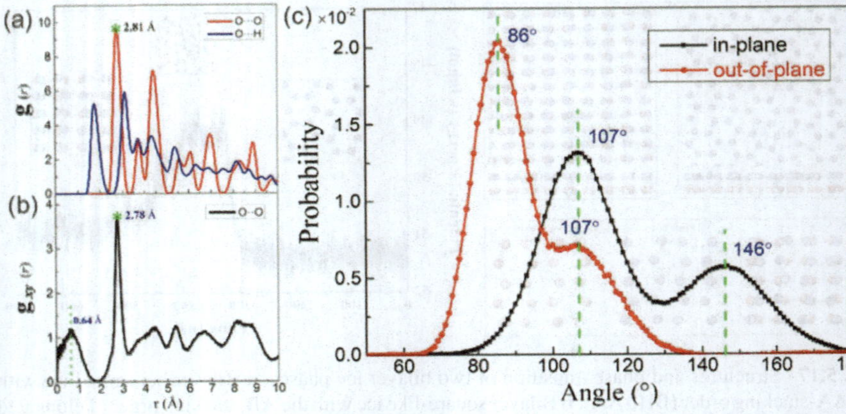

Fig. 5.18 Structural characteristics of BHI-AB from MD simulations. **a** Oxygen-oxygen and oxygen-hydrogen radial distribution functions. **b** Lateral oxygen-oxygen radial distribution function. **c** The in-plane and out-of-plane O···O···O angle distributions in the entire structure of BHI-AB

energy and the diffusion constant increases up to 2×10^{-7} cm^2/s [11], whereas the BHI-AA has the lowest potential energy.

To gain more insight into the structural feature of BHI-AB, we compute the radial distribution function (RDF) and O···O···O angle distributions of BHI-AB. In Fig. 5.18a, the first sharp peak in oxygen-oxygen RDF is located at ~ 2.81 Å, corresponding to the first neighboring distance among oxygen atoms in BHI-AB. The well-separated peaks and valleys indicate that BHI-AB has a long-range crystalline order. In Fig. 5.18b, the oxygen-oxygen lateral RDF also exhibits a sharp peak at ~2.78 Å. Moreover, the first peak in the lateral RDF is located at ~0.64 Å, indicating the staggered offset of the oxygen atoms in the AB-stacked elongated hexagonal rings. Figure 5.18c illustrates the in plane and out-of-plane O···O···O angle distributions in snapshots of BHI-AB in MD simulations within 100 ns equilibrium run. For the in-plane angle distribution, the two peaks are located at 107° and 146 Å, with the ratio of the two corresponding peak values being 2:1. The two angles correspond to the two inner angles of elongated hexagonal ring in BHI-AB. The out-of-plane angle distribution demonstrates that its first peak is located at about 86° with a shoulder being located at 107 Å, indicating that the interlayer hydrogen bonds are not perpendicular to the plane of each monolayer. This feature is consistent with the interlayer distance of BHI-AB being smaller than that of BHI-AA, as the AB-stacking order for the two monolayers allows a small offset of oxygen atoms in the two monolayers as seen from the top view.

We also performed extensive MD simulations to obtain the temperature *versus* nanoslit width (T-h) phase diagram that includes BHI-AA and BHI-AB phases at a given lateral pressure of 450 MPa. This phase diagram plotted in Fig. 5.19 can give an intuitive perspective for the two bilayer solid phases and a bilayer liquid water. The solid-liquid phase boundaries are determined by using the two-phase

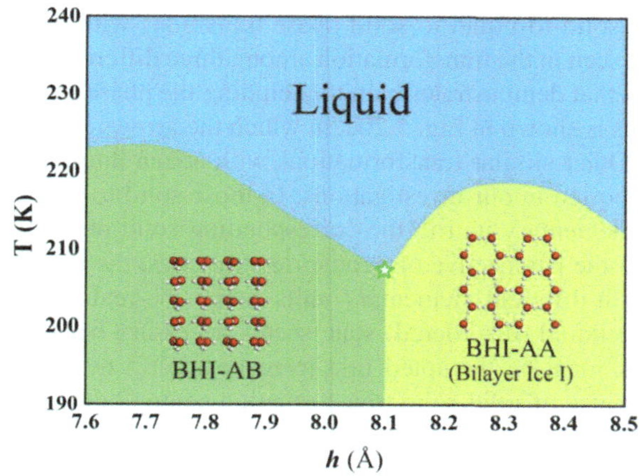

Fig. 5.19 Phase diagram of BHI-AB and BHI-AA in the *T*-*h* plane at the given lateral pressure of 450 MPa. The green open star is the triple point

coexistence simulation method. A triple point (green star in Fig. 5.19) for the three-phase coexistence (BHI-AB/liquid/BHI-AA) is located within the region of nanoslit width 8.0–8.2 Å and of temperature 205–210 K. Such a solid-liquid-solid triple point exhibits some resemblance to that shown in the pressure-temperature phase diagram for the bulk ice I-water-ice III phases. when the temperature is between 205 and 220 K, there exists a liquid region between the two solid phases, corresponding to the solid-to-liquid-to-solid phase transition shown in Fig. 5.17d.

In addition, we also found the phase transition from BHI-AB to BL-VHDI during the compression process. Figure 5.20a shows the abrupt increase in area density, indicating the first-order phase transition from BHI-AB to BL-VHDI. Figure 5.20b shows that transformation from BHI-AB to BL-VHDI also exhibits the Oswald staging phenomenon, namely, an unstable intermediate liquid state arises during the solid-solid phase transition [11].

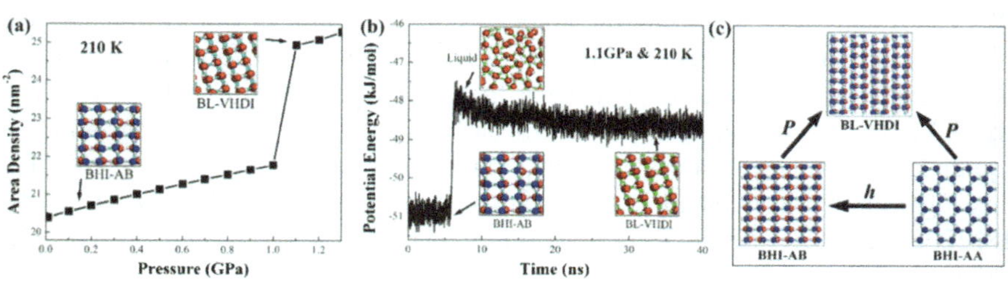

Fig. 5.20 Phase transitions among different bilayer ice phases. **a** Area density as a function of lateral pressure during the compression process of BHI-AB. **b** Time-dependent potential energy (per molecule) during the transformation from BHI-AB to BL-VHDI. **c** Schematic relationship of phase transition among three different bilayer ice phases

Note that the solid-to-liquid-to-solid phase transitions with the Oswald staging phenomenon are seen in the transformation among three different bilayer ice phases. A schematic plot that demonstrates path to including the phase transition among the three solid phases is shown in Fig. 5.20c, in which the arrows give the conditions of transformation. Other similar transformations with arisen intermediate liquid state has also been reported in our investigations. In those solid-to-liquid-to-solid phase transitions, the elementary unit in the corresponding solid phase is different from each other. From the perspective of geometric structures, the direct transformation among solids with different elementary units cannot be realized in 2D water/ice. The intermediate liquid (disordered) state seems necessary because the hydrogen-bonding network must be disrupted first to reconstruct a new hydrogen-bonding network in the course of solid-solid transformation among bilayer ices.

5.4 Summary

In this chapter, we investigated the phase transitions of bilayer water/ice confined in graphene nanocapillaries. In our MD simulations, several bilayer ice phases were found, including AA stacking order and AB stacking order. BL-ABI is a proton-ordered ice with square-like pattern, which can be transformed from pMSI under ultrahigh lateral pressure and ambient temperature. The phase transition from pMSI to BL-ABI is through crystal-to-amorphous-to-crystal pathway with notable hysteresis-loop in the potential energy during the compression/decompression process, reflecting the compression/tensile limit of the 2D monolayer/bilayer ice. BHI-AB is also a square-like ice phase consists of two hexagonal monolayers, each with elongated hexagonal rings, and the hexagonal rings being slightly distorted with interior angle about 107° and 146°. In the transformations among different bilayer ices, solid-to-liquid-to-solid phase transition behavior (or Oswald staging) can arise due to that the hydrogen-bonding network must be disrupted first to reconstruct a new hydrogen-bonding network in the course of solid-solid transformation. For AA-stacked bilayer ice phases, we found a temperature threshold (~275 K) between BL-VHDI and BL-AAI during the compression simulations of bilayer water. The compression and superheating of BL-VHDI display similar structural evolution; that is, square ice-nanotube arrays will bend first, followed by the transformation into BL-AAI. This structural transformation can be regarded as the buckling failure of square ice-nanotube columns, which is dominated by the high lateral pressure. (The investigation in this chapter has been published in *Phys. Chem. Chem. Phys.*, 2016, **18**, 22039–22046; *J. Chem. Phys.*, 2016, **145**, 054704; *J. Phys. Chem. C*, 2018, **122**, 6704–6712.)

References

1. Soper AK (2002) Water and ice. Science 297:1288–1289
2. Zheng Q, Durben DJ, Wolf GH, Angell CA (1991) Liquids at large negative pressures: water at the homogeneous nucleation limit. Science 254:829–832
3. Poole PH, Sciortino F, Essmann U, Stanley HE (1992) Phase behaviour of metastable water. Nature 360:324–328
4. Hummer G, Rasaiah JC, Noworyta JP (2001) Water conduction through the hydrophobic channel of a carbon nanotube. Nature 414:188
5. Koga K, Gao GT, Tanaka H, Zeng XC (2001) Formation of ordered ice nanotubes inside carbon nanotubes. Nature 412:802–805
6. Huang Y, Zhu C, Wang L et al (2016) A new phase diagram of water under negative pressure: the rise of the lowest-density clathrate s-III. Sci Adv 2:e1501010
7. Algara-Siller G, Lehtinen O, Wang FC, Nair RR, Kaiser U, Wu HA, Geim AK, Grigorieva IV (2015) Square Ice in graphene nanocapillaries. Nature 519:443–445
8. Russo J, Romano F, Tanaka H (2014) New metastable form of ice and its role in the homogeneous crystallization of water. Nat Mater 13:733–739
9. Debenedetti PG, Stanley HE (2003) Supercooled and glassy water. Phys Today 56:40–46
10. Johnston JC, Kastelowitz N, Molinero V (2010) Liquid to quasicrystal transition in bilayer water. J Chem Phys 133:154516
11. Bai J, Zeng XC (2012) Polymorphism and polyamorphism in bilayer water confined to slit nanopore under high pressure. Proc Natl Acad Sci USA 109:21240–21245
12. Zhao WH, Wang L, Bai J, Yuan LF, Yang JL, Zeng XC (2014) Highly confined water: two-dimensional ice, amorphous ice, and clathrate hydrates. Acc Chem Res 47:2505–2513
13. Koga K, Zeng XC, Tanaka H (1997) Freezing of confined water: a bilayer ice phase in hydrophobic nanopores. Phys Rev Lett 79:5262
14. Slovak J, Tanaka H, Koga K, Zeng XC (2003) Computer simulation of bilayer ice structures and thermodynamics. Phys A 319:163
15. Koga K, Tanaka H, Zeng XC (2000) First order transition in confined water between high-density liquid and low-density amorphous phases. Nature 408:564–567
16. Bai J, Zeng XC, Koga K, Tanaka H (2003) Formation of quasi two-dimensional bilayer ice in hydrophobic slits: a possible candidate for ice XIII? Mol Simul 29:619–626
17. Han SH, Choi MY, Kumar P, Stanley HE (2010) Phase transitions in confined water nanofilms. Nat Phys 6:685–689
18. Chen J, Schusteritsch G, Pickard CJ, Salzmann CG, Michaelides A (2016) Two dimensional ice from first principles: structures and phase transitions. Phys Rev Lett 116:025501
19. Corsetti F, Zubeltzu J, Artacho E (2016) Enhanced configurational entropy in high-density nanoconfined bilayer ice. Phys Rev Lett 116:085901
20. Koga K, Tanaka H (2005) Phase diagram of water between hydrophobic surfaces. J Chem Phys 122:104711
21. Zangi R, Mark AE (2003) Monolayer ice. Phys Rev Lett 91:025502
22. Zhu YB, Wang FC, Bai J, Zeng XC, Wu HA (2015) Compression limit of two-dimensional water constrained in graphene nanocapillaries. ACS Nano 9:12197–12204
23. Schmidt M, Lowen H (1997) Phase diagram of hard spheres confined between two parallel plates. Phys Rev E Stat Phys Plasmas Fluids Relat Interdiscip Top 55:7228
24. Ayappa KG, Ghatak C (2002) The structure of frozen phases in slit nanopores: a grand canonical Monte Carlo study. J Chem Phys 117:5373
25. Zhu YB, Wang FC, Bai J, Zeng XC, Wu HA (2016) Formation of trilayer ices in graphene nanocapillaries under high lateral pressure. J Phys Chem C 120:8109–8115
26. Jia M, Zhao WH, Yuan LF (2014) New hexagonal-rhombic trilayer ice structure confined between hydrophobic plates. Chin J Chem Phys 27:15–19

Chapter 6
Trilayer Ice in Graphene Nanocapillaries

In MD simulations, we found three different trilayer ice phases, namely, TL-ABAI, TL-ABA, and TL-AAAI, respectively. In particular, TL-ABAI exhibits a square lattice in each layer, which can be viewed as a bilayer clathrate hydrate with water molecules in the middle layer serving as the guest molecules. The phase transition and dynamic behavior of trilayer water in graphene nanocapillaries was systematically studied.

6.1 Brief Introduction

In a constrained high-pressure environment, water can form new solid phases that not seen in the open air [1]. The special feature of confined water stems not only from the intrinsic hydrogen-bonding network but also from the inhomogeneity of confined liquid, particularly in the nanoscale confinement [2]. A better understanding of the effects of confinement and lateral pressure on the phase behavior of water will further the advancement of fundamental research of water [2–6]. In this chapter, we mainly focused on the phase behavior of trilayer water in graphene nanocapillaries under high lateral pressure.

Note that, several trilayer phases of water confined between two parallel hydrophobic surfaces had been predicted from previous studies [2, 7–12]. In MD simulations, Giovambattista et al. found that the middle water layer can behave quite differently from the two outer layers next to the slit walls. Interestingly, the trilayer water is characterized by a middle liquid-like layer sandwiched between two crystal-like layers [7]. Kumar et al. obtained a trilayer ice in which two outer layers are flat while the middle layer is puckered [9]. Flat and puckered trilayer ice phases were also found by Kastelowitz et al. [8] in MD simulations. Zangi and Mark

Y. Zhu, *Phase Behavior of Two-Dimensional Water Confined in Graphene Nanocapillaries*, Springer Theses,
https://doi.org/10.1007/978-981-15-7957-8_6

found a trilayer ice during the freezing simulations of trilayer water under a uniform external electric field in the lateral direction [13]. Recent experiments demonstrated that 2D water in graphene nanocapillaries can transform into monolayer solid phase, in which the van der Waals pressure (GPa magnitude) is the key factor [1]. MD simulations and DFT calculations also revealed rich 2D crystalline and amorphous phases of water in graphene nanocapillaries [14–17]. These studies mainly reported the monolayer square-like ice and the stacking order of bilayer ice. To our knowledge, a comprehensive simulation study of trilayer crystalline/amorphous phases and associated solid-to-liquid-to-solid phase transitions for water confined in graphene nanocapillaries is still lacking.

In this chapter, we mainly concerned the phase behavior of trilayer water confined in graphene nanocapillaries. Three different trilayer phases were obtained from our MD simulations. In particular, the trilayer square ice with the ABA stacking order was reported. Phase behavior of trilayer water in graphene nanocapillaries was investigated in detail.

6.2 Model and Methods

In this chapter, we mainly used the main model in Fig. 2.1 to investigate the structures and phase behavior of trilayer water constrained in graphene nanocapillaries. More details of the MD simulations can be found in Sect. 2.2. In this chapter, the channel width (h) was mainly set to be 9.0, 10.0, and 11.5 Å. The temperature was fixed at 300 K, and the lateral pressure was in the range from 1.0 to 6.0 GPa. In MD simulations, potential energy (per molecule), oxygen-oxygen radial distribution function, oxygen density profile, and mean-squared displacement were calculated to investigate the phase behavior of trilayer ice phases.

6.3 Results and Discussion

6.3.1 *Structures of Different Trilayer Ices in Graphene Nanocapillaries*

Under different conditions of lateral pressure and channel width, phase transition of 2D water in graphene nanocapillaries can be observed during the compression process. Here, we found the formations of three different solid phases of trilayer water in graphene nanocapillaries from MD simulations. The ABA stacking order structure tends to form in relatively narrow graphene nanocapillaries, whereas the AAA stacking order structure tends to form in relatively wide nanocapillaries. As illustrated in Fig. 6.1, based on the structural features, the three phases are named as TL-ABAI, TL-ABA, and TL-AAAI, respectively. Water molecules are colored in

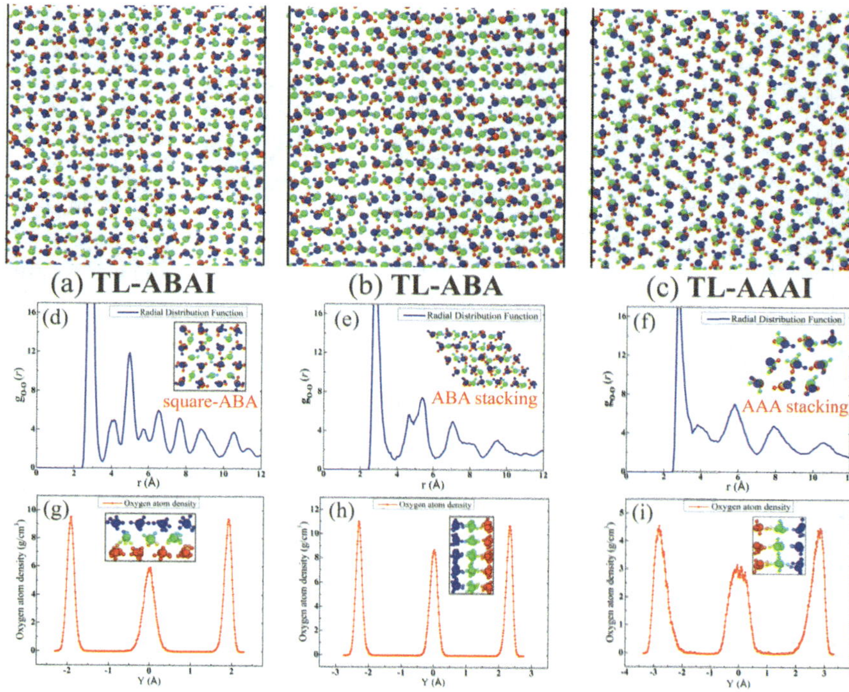

Fig. 6.1 Three different solid phases of trilayer water in graphene nanocapillaries

blue, green, and red to denote three different layers. The insets in Fig. 6.1d–f show the stacking order of each phase. TL-ABAI is the trilayer square ice with the ABA stacking order, in which the third layer (red color) is in registry with the first layer (blue color) while the second (or middle) layer (green color) is staggered with both the first and third layers. The oxygen-oxygen radial distribution function in Fig. 6.1d indicates that oxygen atoms in the TL-ABAI are in an ordered configuration. TL-ABA is the trilayer rhombic amorphous ice with the ABA stacking order, and TL-AAAI is the trilayer triangular ice with the AAA stacking order [12]. Figure 6.1g–i shows the oxygen density profiles of the three phases, indicating that all three phases are very-high-density ices.

TL-ABAI exhibits an inerratic structure, which can be transformed from AA-stacked bilayer triangular ice (BL-AAI) under ultrahigh lateral pressure. Figure 6.2 shows the detailed structure of TL-ABAI. Each water molecule in the middle layer is located in the center of a square prism formed by eight water molecules in the first and third layers. Hydrogen-bonding network in Fig. 6.2a indicates that water molecules in the middle layer could form hydrogen bonds with adjacent layers. As shown in Fig. 6.2b, TL-ABAI can be viewed as a bilayer clathrate hydrate with water

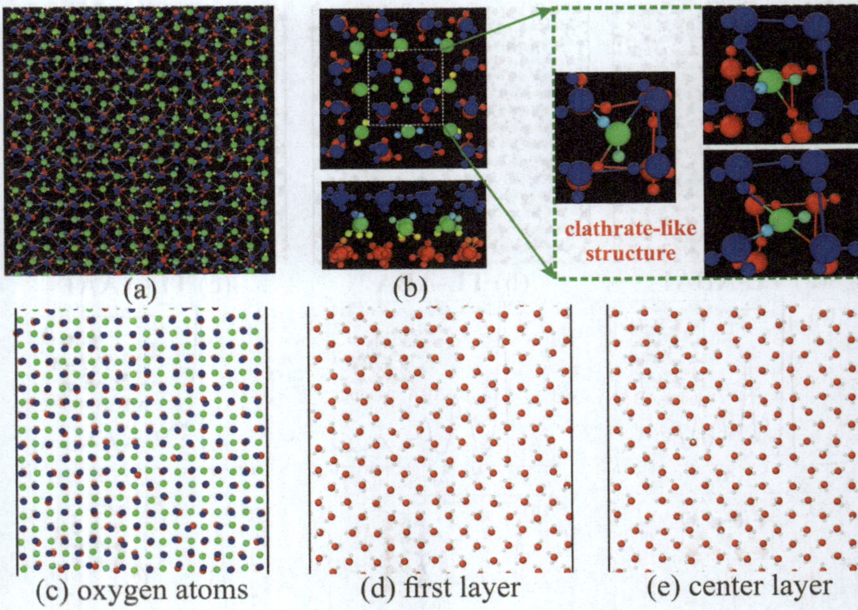

Fig. 6.2 Structure of the trilayer square ice with the ABA stacking order

molecules in the middle layer viewed as the guest molecules. Every four blue water molecules are superimposed on top of another four red water molecules, which form a cube-like structure. Every green water molecule is located at the center of this cube. Figure 6.2c–e illustrates the square pattern in TL-ABAI.

Similarly, TL-ABA can also be regarded as a clathrate-like structure. The main difference between TL-ABAI and TL-ABA is the configuration of oxygen atoms, which is square for TL-ABAI but rhombic for TL-ABA phase. Although the square lattice can be formed in each layer of TL-ABAI, it cannot form a perfect hydrogen-bonding network like the monolayer square-like ice because some hydrogen atoms in one layer form hydrogen bonds with oxygen atoms in the adjacent layers. Note that water molecules in the middle layer are loosely connected with each other.

The transformation from BL-AAI to TL-ABAI is a solid-to-liquid-to-solid phase transition. Figure 6.3a shows the potential energy and the number of water molecules inside the channel during this phase transition. The transformation from TL-AAAI to the ABAB-stacked four-layer amorphous ice is also a solid-to-liquid-to-solid phase transition, and the potential energy in Fig. 6.3b shows a sudden jump at ~4.2 GPa. The five regions in Fig. 6.3b correspond to a liquid phase at low pressure, TL-AAAI, an amorphous trilayer with puckered middle layer, an intermediate liquid state at high pressure, and a four-layer amorphous ice with ABAB stacking order, respectively. The inset in Fig. 6.3b shows that the four-layer amorphous ice with the ABAB

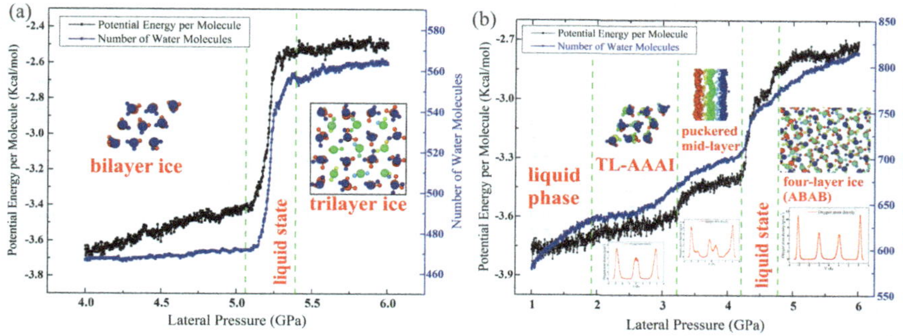

Fig. 6.3 Potential energy (per molecule) during the phase transitions. **a** Phase transition from AA-sacked bilayer triangular ice to TL-ABAI. **b** Phase transition from trilayer liquid water to AAA-stacked trilayer triangular ice and then to four-layer ice with the ABAB stacking order

stacking order exhibit a square-like configuration. Both of the two solid-to-liquid-to-solid phase transitions in Fig. 6.3 display the transformation from AA-stacked structure into AB stacked structure, and one layer of water is increased during the phase transition.

Note that, in both cases of the solid-to-liquid-to-solid phase transition (Fig. 6.3), the Oswald staging phenomenon occurs [6, 18]; namely, an intermediate liquid state becomes stable within a range of lateral pressure between those of the two ice phases. Figure 6.4a, b shows that the structural transformation from triangular to square-like is associated with a sudden increase in the pressure perpendicular to the graphene during the liquid state, due to one more layer of water entering into the channel. Figure 6.4c, d plots the oxygen-oxygen radial distribution function of different phases during the phase transitions. The black lines in Fig. 6.4c, d denote the intermediate liquid state during the solid-to-liquid-to-solid phase transition, which clearly does not show a long-range order. Oxygen-oxygen radial distribution functions of BL-AAI, TL-ABAI, and TL-AAAI exhibit several sharp peaks, while those of intermediate liquid state only shows two low and broad peaks.

6.3.2 *Compression of AAA-Stacked Trilayer Triangular Ice*

During the phase transition from TL-AAAI to ABAB-stacked four-layer amorphous ice, the middle layer of TL-AAAI will be puckered due to the increasing lateral pressure. In our MD simulations, some transient solid structures were observed in the puckered process of middle layer in trilayer ice. As shown in Fig. 6.5, we carried out long-time MD simulation (40 ns) with the lateral pressure ranging from 2.5 to 4.5 GPa. The initial structure of trilayer water in the graphene channel is TL-AAAI. Both of the potential energy and the mean-squared displacement (MSD) show an sudden jump at ~4.05 GPa. The components of MSD in Fig. 6.5b indicate that the diffusion

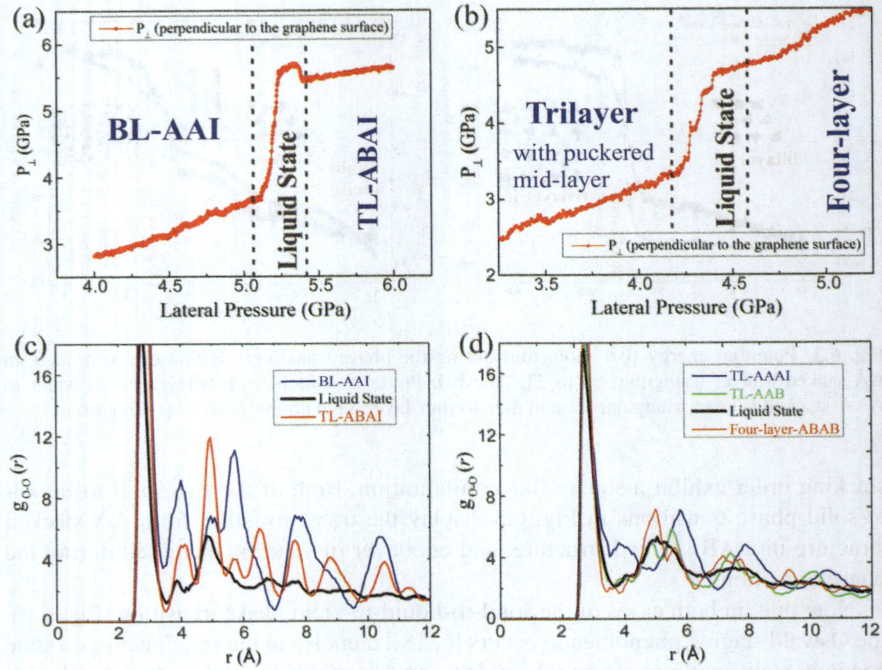

Fig. 6.4 **a**, **b** Pressure in the direction perpendicular to the graphene wall. **c**, **d** Oxygen-oxygen radial distribution functions of different phases

coefficient of trilayer amorphous ice with puckered middle layer is lower than that of TL-AAAI. Figure 6.5c plots several transient structures of trilayer solid phase with the puckered middle layer during this process. In those transient structures, we found an special structure, that is, trilayer amorphous ice with the AAB stacking order (TL-AAB).

As illustrated in Fig. 6.6a, water molecules in TL-AAB shows a parallel line-like configuration, and the middle layer (green color) is in registry with the first layer (blue color) while the third layer (red color) is staggered with both the first and middle layers. The middle layer in TL-AAB is not flat. The oxygen density profile of TL-AAB in Fig. 6.6c exhibits two connected peaks at the position of middle layer, indicating the puckered state of middle layer. However, the structure feature of TL-AAB differs from that of the trilayer ice reported in Kumar et al. [9]. The puckered middle layer does not have a clear zigzag configuration like puckered monolayer ice. As such, the transient structures look more disordered than TL-AAAI.

In TL-AAB, water molecules in each layer exhibit the triangular configuration. The oxygen-oxygen radial distribution function in Fig. 6.6b indicates that TL-AAB is not as regular as TL-ABAI and TL-AAAI. This transient structure can only last for a few nanoseconds. The MSD result in Fig. 6.6d shows that the slope of the MSD curve increases first and then levels off, indicating that TL-AAB changes to another

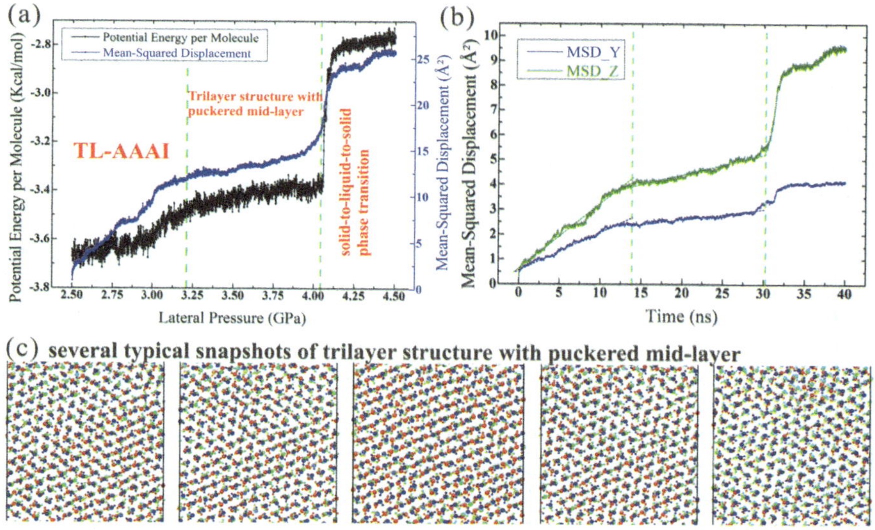

Fig. 6.5 Phase transition during the compression of TL-AAAI (T = 300 K). **a** Potential energy (per molecule) and mean-squared displacement. **b** Mean-squared displacements in the lateral and normal directions. **c** Several transient structures of trilayer phase with the puckered middle layer (3.25 GPa < P_{zz} < 4.0 GPa)

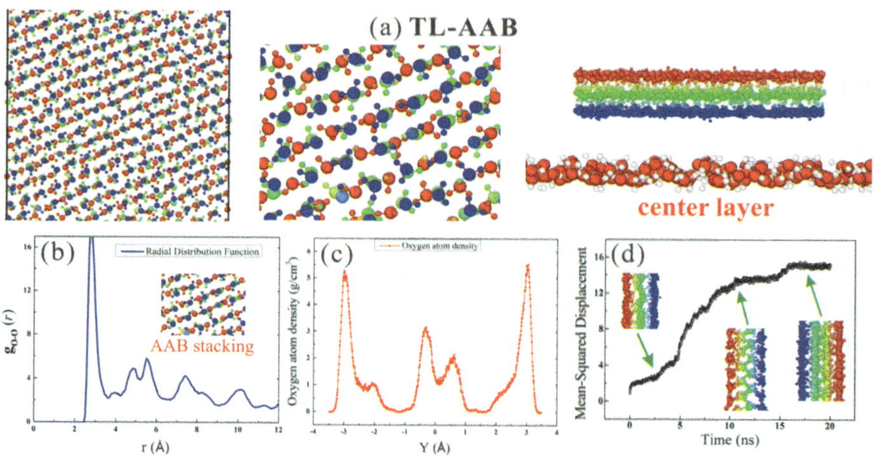

Fig. 6.6 Structural features in AAB-stacked trilayer amorphous ice

structure. Indeed, at the high lateral pressure, TL-AAB turns into another transient structure (see the inset) similar to the trilayer ice reported by Kumar et al. Ultimately, TL-AAAI turns into the ABAB-stacked four-layer amorphous ice.

6.3.3 *Dynamic Behavior of Trilayer Ices*

In Fig. 6.1g–i, we noted that the oxygen density profiles of the three different trilayer phases exhibit a similar feature, that is, the peak value in density of middle layer is much lower than that of two outer layers in contact with the graphene. However, the structural snapshots in Fig. 6.1a–c demonstrate that the numbers of water molecules in each layer are the same. In addition, the width of the middle layer in oxygen density profiles for all three phases is larger than that of two outer layers. Therefore, the behavior of the middle layer should be different from that of the other two outer layers, even though all three layers exhibit similar square or rhombic structures. To further understand the structural differences among the three trilayer phases, we computed the MSDs of oxygen atoms in each layer. The slop of the MSD *versus* time curve is proportional to the diffusion coefficient of water molecules.

MSD curves in Fig. 6.7a–c are the MSD components ($MSD_{\perp}$) of each layer in the direction perpendicular to the graphene wall, while curves in Fig. 6.7d–f show the in-plane MSD components ($MSD_{||}$) of each layer in the direction parallel to the graphene wall. MSD curves are colored in red, green, and blue to denote the MSD data of three different layers in each phase, corresponding the colors of water molecules in different layers. In each phase, MSD curves of two outer layers (red and blue) are nearly the same, but the MSD of the middle layer (green) is notably different from those of the other two layers. In Fig. 6.7a, d, the MSD components of each layer in TL-ABAI show a step-like change and then level off. The step-like changes are closely correlated for the three layers, and these changes are mainly due to vacancies in TL-ABAI. As shown in Fig. 6.8, the migration of the vacancies results in a sudden jump in MSD curves. When the vacancies disappear in the channel, the

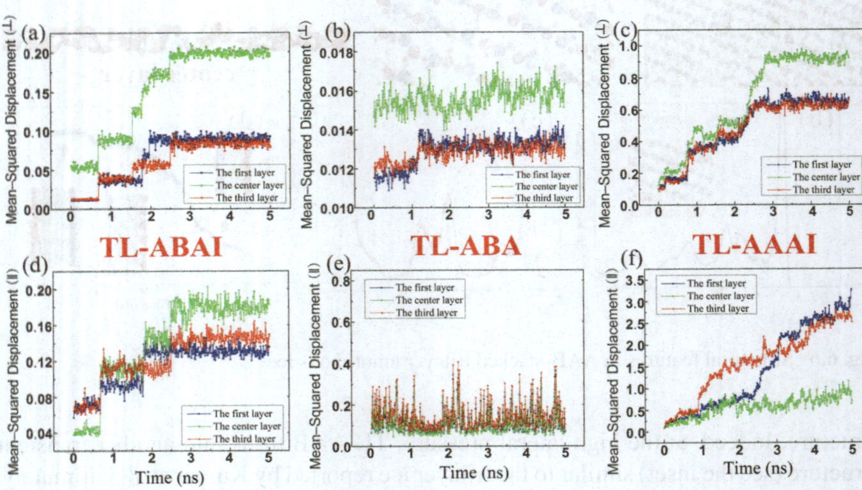

Fig. 6.7 MSDs of each layer in three trilayer ice phases

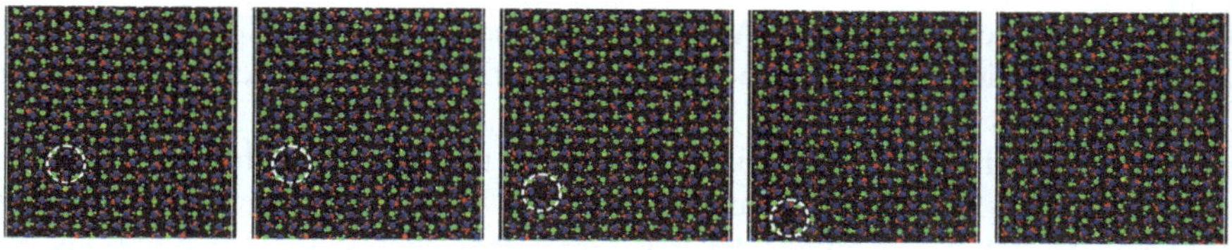

Fig. 6.8 The migration of vacancies in TL-ABAI

MSD curves level off. Both $MSD_{\perp}$ and $MSD_{||}$ of the middle layer are higher than those of the other two layers. For TL-ABA, the $MSD_{\perp}$ of center layer is a bit larger in Fig. 6.7b, and $MSD_{||}$ values of three layers in Fig. 6.7e are nearly the same. For TL-AAAI, unlike the other two trilayer phases, the $MSD_{||}$ of the middle layer is lower than that of two outer layers. The slope of three $MSD_{||}$ curves in Fig. 6.7f suggests that water molecules in the two outer layers move faster. This diffusion behavior of water molecules close to the surface and away from the surface resembles that of simple liquids [9, 19].

MSD curves of the three solid phases under ambient temperature are shown in Fig. 6.9, in which the MSD slopes of TL-ABAI and TL-ABA are close to zero while TL-AAAI exhibits a higher slope in MSD. Based on the MSD curve, we obtained the diffusion coefficient of water molecules in TL-AAAI is ~3.4×10^{-8} cm^2/s, which should be attributed to that the 300 K is close to the melting point of TL-AAAI.

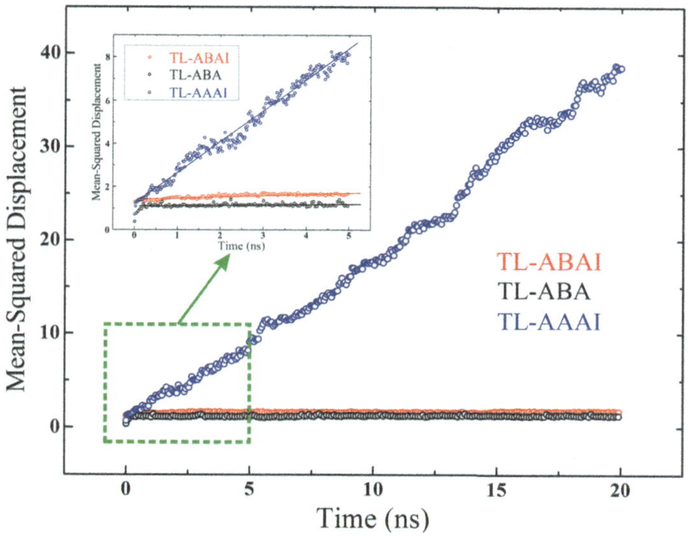

Fig. 6.9 MSD results of three different trilayer ice phases under ambient temperature

We also calculated MSD of TL-AAAI at 240 K and 280 K, and the corresponding diffusion coefficients were ~1.0×10^{-10} cm^2/s and ~1.0×10^{-9} cm^2/s, indicating that TL-AAAI is a solid phase.

The results of MSD in Fig. 6.7 indicate that the dynamic behavior of middle layer is different from the two outer layers close to the graphene surface. For the three phases, $MSD_{\perp}$ of middle layer is larger than that of the other two layers, indicating that the vibration of water molecules in the middle layer in the direction perpendicular to graphene is stronger than that in the other two layers. This result is also corresponding to the feature of oxygen density profiles in Fig. 6.1g–i, where the width of middle layer is wider. Compared to TL-AAAI, the three $MSD_{\|}$ curves are closer to one another for TL-ABAI and TL-ABA, which should attributed to the high-density ABA stacking structure under ultrahigh lateral pressure. The results of $MSD_{\|}$ of TL-AAAI indicate that water molecules in two outer layers move faster than those in middle layer. The MSD results of three layers in each phase display a synchronous feature, reflecting that there are abundant hydrogen bonds between adjacent layers. The motion of each layer is interactive and correlative.

6.4 Summary

In this chapter, we reported three solid phases of trilayer water constrained in graphene nanocapillaries under ultrahigh lateral pressure and ambient temperature. TL-ABAI can be regarded as a clathrate-like inerratic structure with a square lattice in each layer. With increasing lateral pressure, two solid-to-liquid-to-solid phase transitions were observed with the structural transformation from triangular to square-like pattern. For all three phases, the oxygen density profiles show that the peak of the middle layer is lower than those of the other two layers while the width of the middle layer is wider. MSD results show that the middle layer is quite different from the other two outer layers. For TL-AAAI, we found that water molecules in the layer next to graphene move faster than those in the middle layer. (The investigation in this chapter has been published in *J. Phys. Chem. C*, 2016, **120**, 8109–8115.)

References

1. Algara-Siller G, Lehtinen O, Wang FC, Nair RR, Kaiser U, Wu HA, Geim AK, Grigorieva IV (2015) Square ice in graphene nanocapillaries. Nature 519:443–445
2. Zhao WH, Wang L, Bai J, Yuan LF, Yang JL, Zeng XC (2014) Highly confined water: two-dimensional ice, amorphous ice, and clathrate hydrates. Acc Chem Res 47:2505–2513
3. Poole PH, Sciortino F, Essmann U, Stanley HE (1992) Phase behaviour of metastable water. Nature 360:324–328
4. Sciortino F, Essmann U, Stanley HE, Hemmati M, Shao J, Wolf GH, Angell CA (1995) Crystal stability limits at positive and negative pressures and the crystal-to-amorphous transition. Phys Rev E: Stat Phys Plasmas Fluids Relat Interdiscip Top 52:6484–6491

5. Debenedetti PG, Stanley HE (2003) Supercooled and glassy water. Phys Today 56:40–46
6. Bai J, Zeng XC (2012) Polymorphism and polyamorphism in bilayer water confined to slit nanopore under high pressure. Proc Natl Acad Sci USA 109:21240–21245
7. Giovambattista N, Rossky PJ, Debenedetti PG (2009) Phase transitions induced by nanoconfinement in liquid water. Phys Rev Lett 102:050603
8. Kastelowitz N, Johnston JC, Molinero V (2010) The anomalously high melting temperature of bilayer ice. J Chem Phys 132:124511
9. Kumar P, Buldyrev SV, Starr FW, Giovambattista N, Stanley HE (2005) Thermdynamics, structure, and dynamics of water confined between hydrophobic plates. Phys Rev E 72:051503
10. Jia M, Zhao WH, Yuan LF (2014) New hexagonal-rhombic trilayer ice structure confined between hydrophobic plates. Chin J Chem Phys 27:15–19
11. Qiu H, Zeng XC, Guo W (2015) Water in inhomogeneous nanoconfinement: coexistence of multilayered liquid and transition to ice nanoribbons. ACS Nano 9:9877–9884
12. Zhu YB, Wang FC, Bai J, Zeng XC, Wu HA (2015) Compression limit of two-dimensional water constrained in graphene nanocapillaries. ACS Nano 9:12197–12204
13. Zangi R, Mark AE (2004) Electrofreezing of confined water. J Chem Phys 120:7123–7130
14. Chen J, Schusteritsch G, Pickard CJ, Salzmann CG, Michaelides A (2016) Two dimensional ice from first principles: structures and phase transitions. Phys Rev Lett 116:025501
15. Corsetti F, Zubeltzu J, Artacho E (2016) Enhanced configurational entropy in high-density nanoconfined bilayer ice. Phys Rev Lett 116:085901
16. Mario MSF, Neek-Amal M, Peeters FM (2015) AA-stacked bilayer square ice between graphene layers. Phys Rev B: Condens Matter Mater Phys 92:245428
17. Corsetti F, Matthews P, Artacho E (2016) Structural and configurational properties of nanoconfined monolayer ice from first principles. Sci Rep 6:18651
18. Bai J, Zeng XC, Koga K, Tanaka H (2003) Formation of quasi two-dimensional bilayer ice in hydrophobic slits: a possible candidate for ice XIII? Mol Simul 29:619–626
19. Scheidler P, Kob W, Binder K (2002) Cooperative motion and growing length scales in supercooled confined liquids. Europhys Lett 59:701–707

Chapter 7
Compression Limit of 2D Water in Graphene Nanocapillaries

When water is constrained between two sheets of graphene, it becomes a 2D liquid and then is turned into an intriguing monolayer solid with a square pattern due to the high van der Waals pressure. From a mechanics point of view, this liquid-to-solid transformation characterizes the compression limit (or metastability limit) of 2D liquid. In this chapter, we systemically investigated the compression limit of 2D water confined in graphene nanocapillaries and the associated phase diagram was obtained.

7.1 Brief Introduction

An abundance of water on Earth renders its involvement in a vast number of processes that have important relevance to many fields of science [1]. The phase behavior of water is a topic of perpetual interest due to its remarkable anomalous properties and important implications. Less obvious but undoubtedly ubiquitous is the phase behavior of water confined in highly constrained nanoscale spaces. In the constrained high-pressure environment, water can form new phases not seen in open air [2]. The metastability limit extension of the phase diagram of liquid water at high compression/tensile pressure or high degree of superheating/supercooling provides new features that manifest themselves in the equilibrium properties of water [3–6]. Evaluation of the tensile/compression limit of a solid under conditions of tension or compression is often performed to provide mechanical properties that are critical for structure design and assessment. The effects of the constraint and associated high pressure on the mechanical stability of water are yet to be comprehensively studied [7–26].

Y. Zhu, *Phase Behavior of Two-Dimensional Water Confined in Graphene Nanocapillaries*, Springer Theses,
https://doi.org/10.1007/978-981-15-7957-8_7

In a recent experimental study [2], Algara-Siller et al. discovered monolayer square-like ice between two graphene sheets by transmission electron microscopy. In this experiment, a water nanodroplet was initially placed on one graphene sheet, and another graphene sheet was then placed on the top, covering the water droplet. The strong graphene-graphene interaction makes the two graphene sheets come together at the margin, and the van der Waals (vdW) pressure squeezes some water out of the channel while some others are trapped inside the graphene pockets and form few-layer ice. Meanwhile, the complementary MD simulations (with a helium bubble and water nanodroplet) in Ref. [2] indicated that the lateral pressure can simulate and characterize the vdW pressure in the graphene nanocapillaries. The monolayer square-like ice observed in the experiment is consistent with some previous simulation studies [19, 22–24]. In turn, the experiment can be also viewed as a demonstration of the compression limit of the 2D water. Indeed, water constrained in graphene nanocapillaries differs from that in carbon nanotubes in that the high vdW pressure becomes an important factor to induce the square-like ice structures [2]. This is because for water constrained in the graphene pockets the vdW force or London dispersion force [27, 28] can be as high as 1 GPa. It is of both fundamental and technological importance to investigate the mechanical stability of highly constrained water, including the compression limit of constrained water in graphene nanocapillaries, the lateral pressure-dependent phase diagram, the pore-width dependence of the freezing and melting, as well as the lateral pressure conditions for gradual and abrupt freezing and melting.

7.2 Model and Methods

In this chapter, we mainly used the main model in Fig. 2.1 to investigate the structures and phase behavior of trilayer water constrained in graphene nanocapillaries. More details of the MD simulations can be found in Sect. 2.2. In this chapter, the channel width (h) was ranging from 6.0 to 11.6 Å, allowing one to three monolayers of water to be accommodated within the graphene channel. The range of lateral pressure (P_{zz}) was from 0.1 GPa to 3.5 GPa. Temperature was fixed at 300 K. The discussion on the selection of water model in MD simulations could be found in Sect. 2.2.

7.3 Results and Discussion

7.3.1 Monolayer Square-Like Ice and First-Order Phase Transition

In the first series of MD simulations, the temperature was fixed at 300 K and the channel width was 7.0 Å, while the lateral pressure was changed in a step of 0.05–0.1 GPa during the compression and decompression processes. Figure 7.1 shows the change of potential energy (per molecule) during the pressurization and depressurization processes. During the compression process, potential energy decreases gradually and then drops suddenly by about 0.25 kcal/mol at 600 MPa. This obvious energy drop within a pressure range of 50 MPa is mainly due to the formation of monolayer solid with ordered hydrogen-bonding network from initial liquid state, which gives the compression limit of monolayer water. In the reversed decompression process, potential energy increases gradually and then rapidly jumps at 500 MPa, following the path of pressurization process. This large jump in potential energy is mainly due to the hydrogen-bonding network was destroyed and the monolayer solid transformed into liquid water, which gives the tension limit of monolayer ice. The sudden change in potential energy and associated large hysteresis loop indicate that the transformation between monolayer water and monolayer square-like ice is a first-order phase transition [13]. In particular, in the range of 500–600 MPa, the compression/decompression process triggers the freezing/melting transition, giving the compression/tension limit of 2D monolayer water/ice. Moreover, this hysteresis

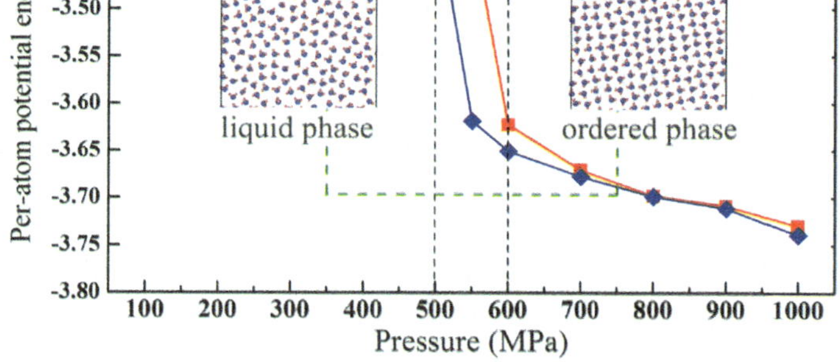

Fig. 7.1 Potential energy (per molecule) as a function of lateral pressure during the compression and decompression processes. Two insets show the liquid state and ordered monolayer phase

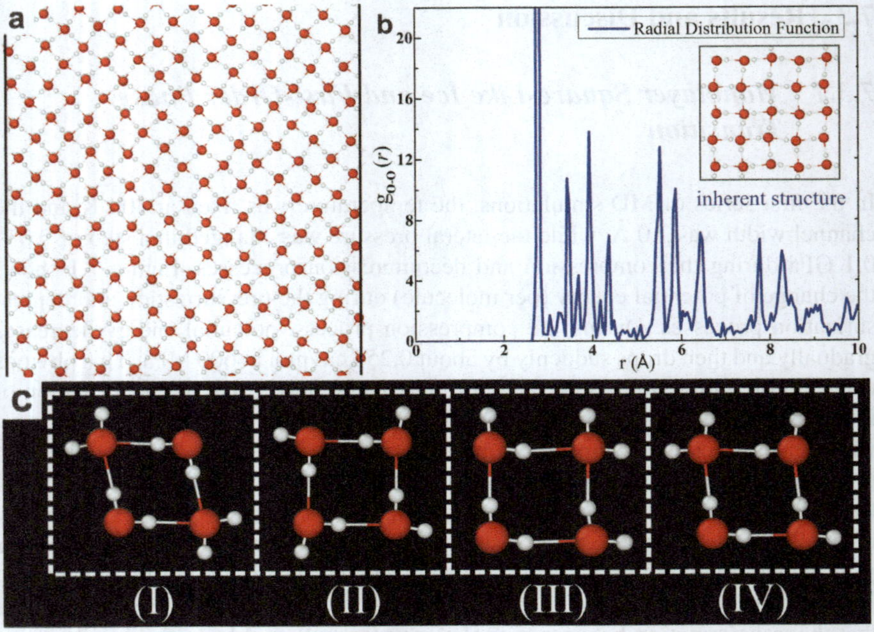

Fig. 7.2 Snapshot of flat monolayer square-like ice (fMSI) and its structural feature

loop in potential energy was also observed in other MD simulations when the channel width is smaller than 7.6 Å. The initial state of water molecules in graphene channel driven by lateral pressure is liquid monolayer. When the lateral pressure is beyond the corresponding compression limit, monolayer water transformed into monolayer square-like ice.

Figures 7.2 and 7.3 illustrate the typical snapshots of monolayer ice structures obtained from our MD simulations. When the channel width is 6.0 Å, the 2D ice is flat monolayer square-like ice (fMSI). When the channel width is 7.0 Å, the 2D ice is puckered monolayer square-like ice (pMSI). In Fig. 7.3, water molecules are colored by the coordinate values in the direction perpendicular to the graphene wall. In previous studies, similar phases of monolayer ice were also found in the confinement composed by two parallel plates [7, 8, 15, 21, 22]. Both the two monolayer structures satisfy the ice rule that each water molecule is hydrogen-bonded to its four nearest neighbors [16, 23], with every water molecule being a double donor and a double acceptor of hydrogen bonds [12]. Each node of the square-like network is occupied by an oxygen atom except some point defects. The two monolayer ice are not strictly square because the H–O–H bond angle (104.52°) is not exactly a right angle.

To further shed light on the structural difference between fMSI and pMSI, oxygen-oxygen radial distribution function ($\mathbf{g}_{O\text{-}O}(r)$) and structural details are shown in Figs. 7.2 and 7.3. For fMSI, the oxygen density profile shows only one peak, indicating that all water molecules in fMSI are nearly in the same plane. For pMSI, the

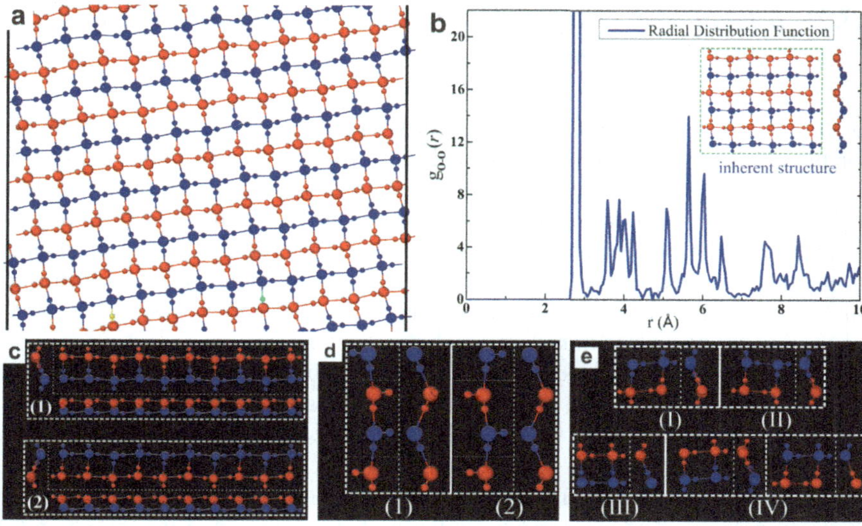

Fig. 7.3 Snapshot of puckered monolayer square-like ice (pMSI) and its structural feature

oxygen density profile exhibits two connected peaks as this phase is a puckered structure with net polarization [7, 22, 23]. The red and blue hydrogen bonds in Fig. 7.3a form ridges and valleys of the puckered structure, respectively. The zigzag rows crossing over the channel show that the hydrogen bonds are tilted with respect to the graphene surface, indicating that a number of hydrogen bonds are nearly parallel to the graphene surface while others form an angle with the surface.

When the lateral pressure is 1.0 GPa, the numbers of water molecules for fMSI and pMSI in the graphene channel were about 200 and 220, respectively. If we further increased the lateral pressure, the number of water molecules in the channel would reach a stable value for $h = 6.0$ Å, while for $h = 7.0$ Å, the number of water molecules in the channel would continue to increase because pMSI will transform into a bilayer amorphous phase. The oxygen-oxygen radial distribution functions in Figs. 7.2b and 7.3b indicate the two monolayer phases exhibit long-range order. The first sharp peak in $g_{O\text{-}O}(r)$s for both fMSI and pMSI is located at 2.79 ± 0.06 Å [2], and some low peaks are seen at the location of multiples. Figures 7.2c and 7.3c–e show the structural details of the two monolayer square-like phases. Monolayer square-like ice is composed of four different elementary units, and the main configuration is the assembled long-range ordered structure with alternative rhombic unit and square unit. Therefore, monolayer square-like ice displays the slightly-rhomboidal feature in our MD simulations. In Fig. 7.3c–e, the combination of different units can influence the zigzag-like configuration of pMSI.

7.3.2 Bilayer and Trilayer Ice Phases and Continuous-Like Phase Transition

For $h > 7.2$ Å, when the lateral pressure is larger than corresponding compression limit, 2D water in graphene nanocapillaries would form amorphous bilayer and trilayer ice phases. Surprisingly, unlike the monolayer square-like ice, hysteresis loop in potential energy is not found in the phase transitions of bilayer and trilayer ice phases. in Fig. 7.4, potential energy curves of bilayer water/ice are nearly overlapped during the pressurization and depressurization processes, indicating the transformation between bilayer water and bilayer ice in current system ($h = 9.0$ Å, $T = 300$ K) is a continuous-like phase transition.

Based on the stacking order of water molecules, bilayer ice phases can be classified into two types; that is, AA-stacked structure and AB-stacked structure [2], respectively. In AA stacking structure, oxygen atoms in adjacent layers are superimposed on top of the other, whereas in the AB stacking structure, oxygen atoms in the second layer are no longer located above the oxygen sites of the first layer. The bilayer AB-stacked ice phase is similar to that reported by Zangi based on the TIP5P water model [19]. In our MD simulations, under ambient temperature and high lateral pressure, 2D water confined in graphene nanocapillaries can spontaneously form bilayer (AA and AB stacking) and trilayer (AAA and ABA stacking) ice phases. The structures of those bilayer and trilayer phases are shown in Figs. 7.5 and 7.6.

When the channel width is 9.0 Å, as shown in Figs. 7.5b and 7.6a, oxygen atoms in adjacent layers are superimposed on top of one another. Every three neighboring water molecules in the same layer form a triangle, as illustrated by the isometric green dashed lines. This AA-stacked bilayer triangular ice (BL-AAI) is different from

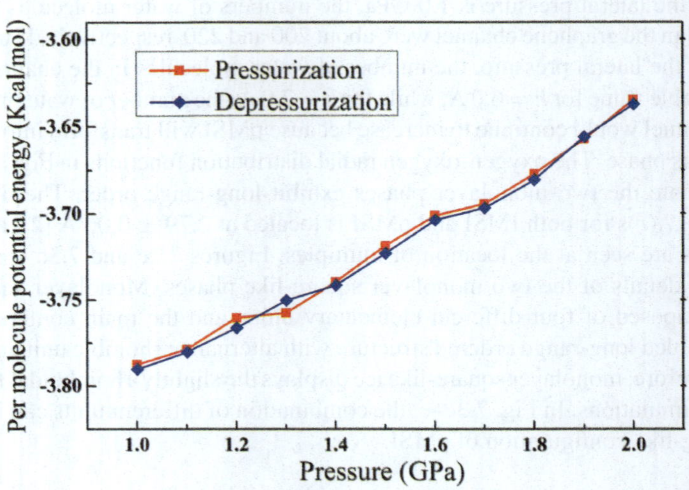

Fig. 7.4 Potential energy of bilayer water/ice ($h = 9.0$ Å, $T = 300$ K) during the pressurization and depressurization processes

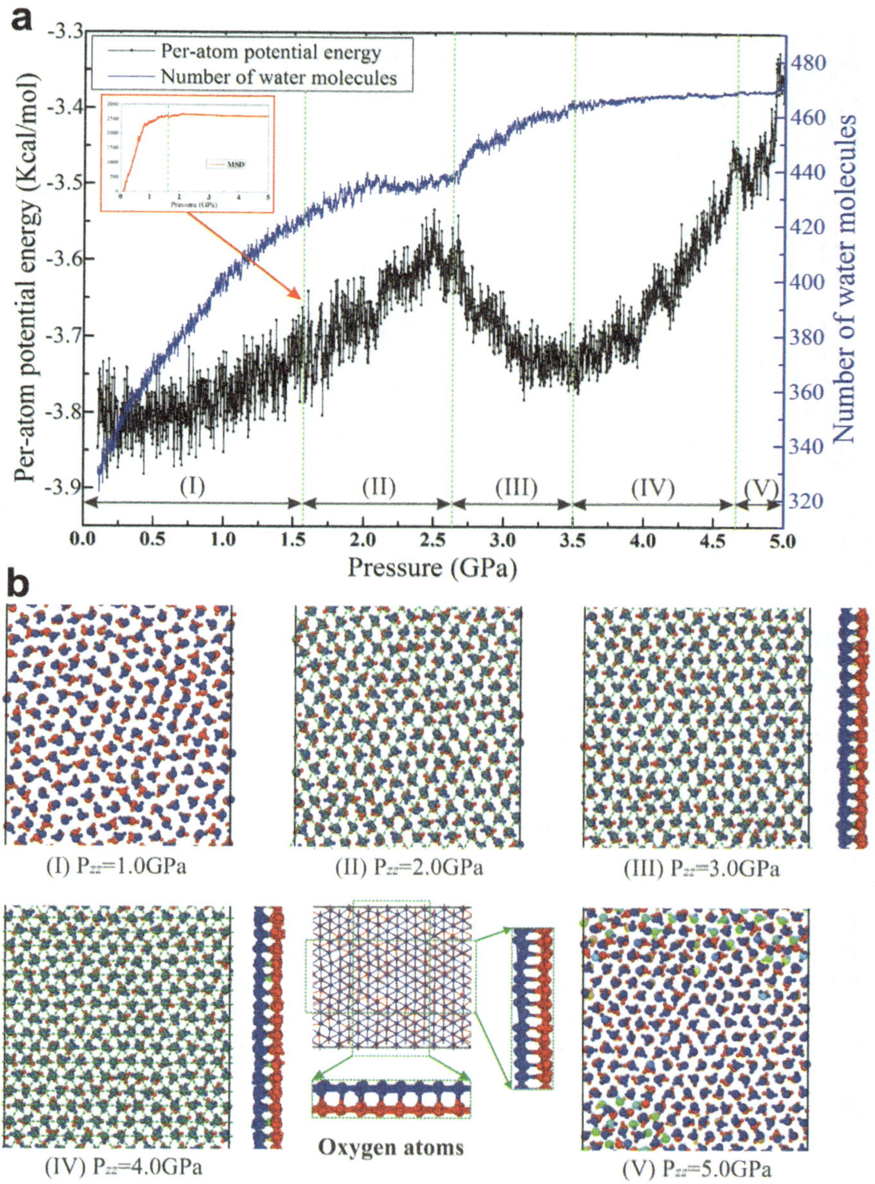

Fig. 7.5 Bilayer AA-stacked triangular ice (BL-AAI) and its structural features. **a** Potential energy per molecule. **b** Structural snapshots under different lateral pressure

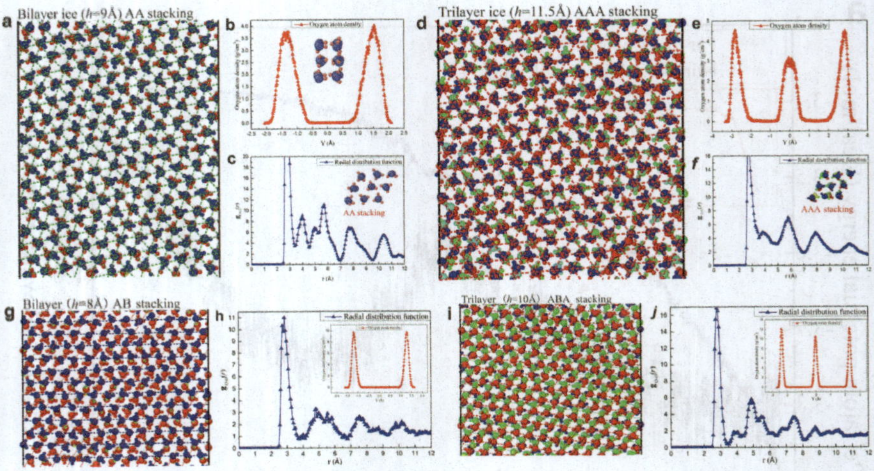

Fig. 7.6 Structures of bilayer and trilayer ice phases in graphene nanocapillaries under ambient temperature. **a–c** Bilayer AA-stacked triangular ice. **d–f** Trilayer AAA-stacked triangular ice. **g, h** Bilayer AB-stacked amorphous ice. **i, j** Trilayer ABA-stacked amorphous ice

the BL-VHDI in previous studies [7, 26]. Figure 7.5a plots the pressure-dependent compression limit of bilayer water under ambient temperature. Typical snapshots in Fig. 7.5b show the structural evolution during the formation of BL-AAI. For this bilayer ice, interestingly, the potential energy (per molecule) is higher than that of corresponding bilayer liquid water. From the front view in Fig. 7.6c, only three hydrogen atoms can be seen for each superimposed water molecules. The fourth hydrogen atom forms an inerratic hydrogen bond with the oxygen atom in the opposing layer. Thus, the interlayer interaction is relatively strong for BL-AAI because of the interlayer hydrogen bonds. These interstratified hydrogen bonds are likely the primary reason for the high stability of the AA-stacking structure. The oxygen-oxygen radial distribution function of BL-AAI in Fig. 7.6c indicates that the configuration of oxygen atoms is regular [19]. The first sharp peak in Fig. 7.6c is located at ~2.79 Å, and several relatively low peaks at a distance of multiples of 2.79 Å. Note that this triangular structure of BL-AAI is different from the bilayer square ice structure imaged from a tunneling electron microscopy experiment [2], although both structures exhibit the AA stacking patterns. There is a possibility that the TIP4P/2005 model cannot capture the bilayer square ice structure or that the experimental bilayer square ice is a result of epitaxial growth out of the monolayer square ice. The latter was likely formed first in the experiment and coexisted with the bilayer and trilayer square ice. (Note also that coexistence of 2D ices with different layers has recently been seen in a computer simulation of confined water under the tip of atomic force microscope [29].) The discrepancy requires further investigation from both theoretical modeling and experimental measurements.

When the channel width is 11.5 Å and lateral pressure is 1.85 GPa, as shown in Fig. 7.6d, 2D water in graphene nanochannel transformed into the AAA-stacked trilayer triangular ice (TL-AAAI). The oxygen density profile in Fig. 7.6e shows three independent peaks and the peak value for middle layer is smaller than that of two outer layers. In TL-AAAI, oxygen atoms in adjacent layers (denoted by blue, green, and red, respectively) are superimposed directly on top of one another. In each layer of TL-AAAI, every three water molecules form a triangle, highlighted by the isometric red dashed lines. In Fig. 7.6f, the first sharp peak of oxygen-oxygen radial distribution function is located at ~2.73 Å. Both the BL-AAI and TL-AAAI have almost the same lattice constant (2.79 ± 0.06 Å) as the monolayer square-like ice. In the compression process, potential energy (per molecule) of TL-AAAI increases gradually, which is similar to that of BL-AAI. When the lateral pressure is beyond 4.0 GPa, the trilayer phase would transform into four-layer phase.

During the compression and decompression processes, hysteresis loop in potential energy is not found for BL-AAI and TL-AAAI, indicating that the compression and tension limits are overlapping for bilayer and trilayer water/ice. This is much different from the first-order phase transition of monolayer square-like ice. Therefore, under ambient temperature and high pressure, the phase transition between bilayer (trilayer) water and ice should be continuous-like. Note that this continuous-like phase transition has been observed in the transition between the bilayer liquid water and very-high-density amorphous bilayer rhombic ice [30]. Instead of using the temperature, here, we show that the pressure can also be used as a controlling parameter to realize the continuous-like phase transition of confined water at room temperature.

Moreover, as shown in Fig. 7.6g–j, we also obtained AB-stacked bilayer (BL-AB) and ABA-stacked trilayer (TL-ABA) amorphous ice. The configurations of BL-AB and TL-ABA are disordered and adjacent layers in the two phase are staggered. For $h = 8.0$ Å in Fig. 7.6g, water molecules in the second layer (red) are staggered with those in the first layer (blue). The structure of BL-AB is similar with the very-high-density bilayer amorphous ice (BL-VHDA$_2$) reported by Bai and Zeng [26]. In each layer of BL-AB, water molecules form rhombic-like pattern, while regular hydrogen-bonding network cannot be formed. BL-AB also exhibits many interlayer hydrogen atoms. The oxygen-oxygen radial distribution function of BL-AB in Fig. 7.6h shows the first sharp peak located at ~2.85 Å. The value of first peak in Fig. 7.6h is lower than that in Fig. 7.6c, indicating that BL-AB is an amorphous phase. When the channel width is 10.0 Å, as shown in Fig. 7.6i, the third water molecules layer (blue) is in registry with the first layer (red), while the second layer (green) is staggered with both the first and third layer. TL-ABA shows some kind of ordering with three layers representing the triangular network, as highlighted by red and green dashed lines. Water molecules in TL-ABA cannot form regular hydrogen-bonding network, and water molecules in middle layer are located at the center of rhombic network form by water molecules in two outer layers. The oxygen-oxygen radial distribution function of TL-ABA in Fig. 7.6j shows the first sharp peak located at ~2.79 Å and two low peaks located at 4.89 and 7.41 Å, respectively. For BL-AB and TL-ABA, the compression limits of corresponding liquid water are 2.55 and 1.95 GPa, respectively, which are higher than those of AA and AAA stacking structures. The peak values in oxygen density profiles of BL-AB and TL-ABA are higher than those in BL-AAI and TL-AAAI, indicating that water molecules are more difficult to enter into the relatively narrow graphene nanocapillaries.

7.3.3 Compression-Limit Phase Diagram

Previous studies demonstrated that the freezing point of confined water depends strongly on channel width [7, 8, 16]. Here, we performed extensive MD simulations to explore the compression-limit phase diagram of 2D water in h-P_{zz} plane. In simulations, the range of channel width was from 6.0 to 11.6 Å, and the range of lateral pressure was from 0.1 to 3.5 GPa. The temperature was fixed at 300 K for all cases. Each point in Fig. 7.7 starts from corresponding liquid state, and MD simulations were carried out to measure the compression limit of 2D liquid water. In this phase diagram, we mainly concerned the monolayer, bilayer, and trilayer confined water.

In general, AB-stacked 2D ice is formed in a narrower channel under higher lateral pressure, while the AA stacking ice is easily obtained at a wider channel and lower lateral pressure. In this chapter (P_{zz} < 3.5 GPa), we mainly illustrated six crystalline or amorphous ice phases in the phase diagram: fMSI (6.0 Å ≤ h ≤ 6.7 Å), pMSI (6.7 Å ≤ h ≤ 7.6 Å), BL-AB (7.2 Å ≤ h ≤ 8.3 Å), BL-AAI (8.3 Å ≤ h ≤ 9.2 Å), TL-ABA (9.2 Å ≤ h ≤ 11.1 Å), and TL-AAAI (11.1 Å ≤ h ≤ 11.6 Å). The P_{zz}-dependent

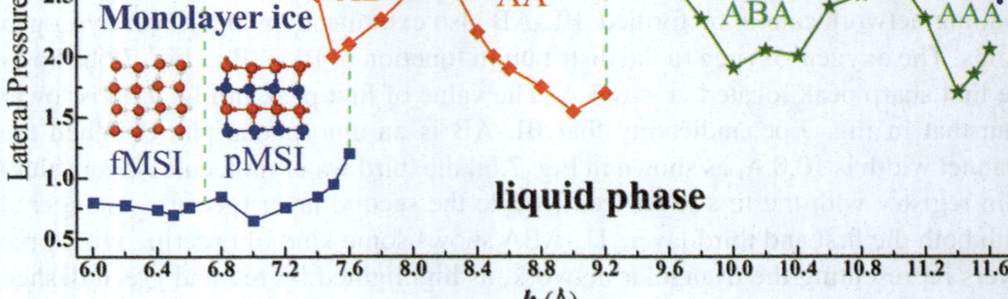

Fig. 7.7 Compression-limit phase diagram in the h-P_{zz} plane of 2D water confined in graphene nanocapillaries under ambient temperature. The insets illustrate the corresponding ice structure. In diagram, the range of lateral pressure is 0.1–3.5 GPa. The star point (9.2 Å, 4.4 GPa) beyond the diagram indicates that BL-AAI can transform into TL-ABAI at ultrahigh lateral pressure. Phase behavior of 2D ice under ultrahigh lateral pressure (>3.5 GPa) was not included in this chapter

compression-limit curves in Fig. 7.7 are obtained based on the compression-limit points at given channel width from MD simulations. Those compression-limit points correspond to the lowest lateral pressures at which an abrupt solid phase is observed in MD simulations. At some h and higher P_{zz}, the monolayer ice (7.2 Å $\leq h \leq$ 7.6 Å) can transform into the bilayer ice, and the bilayer ice (9.0 Å $\leq h \leq$ 9.4 Å) can transform into trilayer ice. Those transformations denote the compression limit of corresponding 2D ice. When the channel width is larger than 11.7 Å, four-layer ice can be formed. Both the lateral pressure and the confinement can influence the phase behavior of 2D water/ice confined in graphene nanocapillaries.

For mono-, bi-, and trilayer ice phases, their compression-limit curves in Fig. 7.7 are concave upward. One simple explanation of this concave upward feature is related to the capillary pressure and vdW pressure in graphene nanocapillaries. For small channel width, the interaction between two sheets of graphene tends to prevent water molecules from entering the nanochannel. Whereas, for large channel width, the nucleation barrier is higher than that for the formation of the 2D ice in narrower slit pore, or the ice would transform into another solid phase under higher lateral pressure. Here, it is assumed that the slope of the compression-limit curve satisfies the Clayperon equation as the freezing curve [8, 16]:

$$\left(\frac{\partial P_{zz}}{\partial h}\right)_T = \frac{(a\Delta P)^\alpha - (a\Delta P)^\beta}{v^\alpha - v^\beta} = \frac{f^\alpha - f^\beta}{v^\alpha - v^\beta},$$

where a is the water-graphene interfacial area per molecule, $\Delta P = P_{yy} - P_{zz}$, v is the volume per molecule, and α and β denote the liquid and solid phases, respectively. At high lateral pressure, $v^\alpha - v^\beta > 0$. When h is less than an optimal value for ice, f^β of the ice is higher than f^α, so $\partial P_{zz}/\partial h < 0$. When h is larger than the optimal value, f^β is smaller than f^α, so $\partial P_{zz}/\partial h > 0$. Thus, the compression-limit curve exhibits a local minimum at the optimal value of channel width.

The slope of ice-ice phase boundaries in Fig. 7.7 is steeper, which should be attributed to that P_{yy} increases faster than P_{zz} when a low-density ice transforms into a high-density ice. With increasing channel width, P_{yy} decreases faster and the lower-density ice tends to form. For example, with increasing P_{zz}, we observed that pMSI maintains the puckered structure first and then transforms into BL-AB. At even higher P_{zz}, TL-AAAI transforms into a four-layer amorphous ice with the middle layers being highly puckered.

7.4 Summary

In this chapter, we systemically investigated the compression limit of 2D liquid water confined in graphene nanocapillaries at ambient temperature. The ranges of channel width and lateral pressure are 6.0 Å $\leq h \leq$ 11.6 Å and 0.1 GPa $\leq P_{zz} \leq$ 3.5 GPa, respectively. Monolayer square-like ice can form regular hydrogen-bonding network, and obvious hysteresis loop in potential energy is found between the compression/decompression process of monolayer water/ice, indicating the first-order phase transition between monolayer water and monolayer square-like ice. Whereas, in the compression/decompression process, potential energy curves are overlapped without hysteresis loop for both bilayer and trilayer water/ice, indicating the continuous-like phase transition between bilayer (and trilayer) ice and corresponding liquid phase. At last, we obtained the compression-limit (phase) diagram of 2D water constrained in graphene nanocapillaries at ambient temperature. (The investigation in this chapter has been published in *ACS Nano*, 2015, **9**, 12197–12204.)

References

1. Brovchenko I, Oleinikova A (2008) Interfacial and confined water. Elsevier, Amsterdam
2. Algara-Siller G, Lehtinen O, Wang FC, Nair RR, Kaiser U, Wu HA, Geim AK, Grigorieva IV (2015) Square ice in graphene nanocapillaries. Nature 519:443–445
3. Zheng Q, Durben DJ, Wolf GH, Angell CA (1991) Liquids at large negative pressures: water at the homogeneous nucleation limit. Science 254:829–832
4. Sciortino F, Essmann U, Stanley HE, Hemmati M, Shao J, Wolf GH, Angell CA (1995) Crystal stability limits at positive and negative pressures and the crystal-to-amorphous transition. Phys Rev E: Stat Phys Plasmas Fluids Relat Interdiscip Top 52:6484–6491
5. Debenedetti PG, Stanley HE (2003) Supercooled and glassy water. Phys Today 56:40–46
6. Poole PH, Sciortino F, Essmann U, Stanley HE (1992) Phase behaviour of metastable water. Nature 360:324–328
7. Zhao WH, Wang L, Bai J, Yuan LF, Yang JL, Zeng XC (2014) Highly confined water: two-dimensional ice, amorphous ice, and clathrate hydrates. Acc Chem Res 47:2505–2513
8. Koga K, Tanaka H (2005) Phase diagram of water between hydrophobic surfaces. J Chem Phys 122:104711
9. Matsumoto M, Saito S, Ohmine I (2002) Molecular dynamics simulation of the ice nucleation and growth process leading to water freezing. Nature 416:409–413
10. Vega C, Sanz E, Abascal JL (2005) The melting temperature of the most common models of water. J Chem Phys 122:114507
11. Kastelowitz N, Johnston JC, Molinero V (2010) The anomalously high melting temperature of bilayer ice. J Chem Phys 132:124511
12. Koga K, Gao GT, Tanaka H, Zeng XC (2001) Formation of ordered ice nanotubes inside carbon nanotubes. Nature 412:802–805
13. Koga K, Tanaka H, Zeng XC (2000) First order transition in confined water between high-density liquid and low-density amorphous phases. Nature 408:564–567
14. Bai J, Wang J, Zeng XC (2006) Multiwalled ice helixes and ice nanotubes. Proc Natl Acad Sci USA 103:19664–19667
15. Bai J, Angell CA, Zeng XC (2010) Guest-free monolayer clathrate and its coexistence with two-dimensional high-density ice. Proc Natl Acad Sci USA 107:5718–5722
16. Takaiwa D, Hatano I, Koga K, Tanaka H (2008) Phase diagram of water in carbon nanotubes. Proc Natl Acad Sci USA 105:39–43
17. Sliwinska-Bartkowiak M, Gras J, Sikorski R, Radhakrishnan R, Gelb L, Gubbins KE (1999) Phase transitions in pores: experimental and simulation studies of melting and freezing. Langmuir 15:6060–6069
18. Kobayashi K, Koshino M, Suenaga K (2011) Atomically resolved images of Ih ice single crystals in the solid phase. Phys Rev Lett 106:206101
19. Zangi R, Mark AE (2003) Bilayer ice and alternate liquid phases of confined water. J Chem Phys 119:1694–1700

20. Tanaka H, Koga K (2005) Formation of ice nanotube with hydrophobic guests inside carbon nanotube. J Chem Phys 123:094706
21. Strauss I, Chan H, Kral P (2014) Ultralong polarization chains induced by ions solvated in confined water monolayers. J Am Chem Soc 136:1170–1173
22. Kaneko T, Bai J, Yasuoka K, Mitsutake A, Zeng XC (2014) Liquid-solid and solid-solid phase transition of monolayer water: high-density rhombic monolayer ice. J Chem Phys 140:184507
23. Zangi R, Mark A (2003) Monolayer ice. Phys Rev Lett 91:025502
24. Zhao WH, Bai J, Yuan LF, Yang J, Zeng XC (2014) Ferroelectric hexagonal and rhombic mono layer ice phases. Chem Sci 5:1757–1764
25. Giovambattista N, Rossky P, Debenedetti P (2009) Phase transitions induced by nanoconfinement in liquid water. Phys Rev Lett 102:050603
26. Bai J, Zeng XC (2012) Polymorphism and polyamorphism in bilayer water con fined to slit nanopore under high pressure. Proc Natl Acad Sci USA 109:21240–21245
27. Rowlinson JS, Widom B (1982) Molecular theory of capillarity (International series of mono graphs on chemistry). Clarendon, Oxford
28. London F (1937) The general theory of molecular forces. Trans Faraday Soc 33:8b–26
29. Qiu H, Zeng XC, Guo W (2015) Water in inhomogeneous nanoconfinement: coexistence of multilayered liquid and transition to ice nanoribbons. ACS Nano 9:9877–9884
30. Han SH, Choi MY, Kumar P, Stanley HE (2010) Phase transitions in confined water nanofilms. Nat Phys 6:685–689

Chapter 8
Summary and Future Work

8.1 Summary

"What is the structure of water" is one of the 125 most challenging scientific questions proposed by *Science* journal. The investigation in phase transitions of confined water is an important branch and front hotspot in the field of water. On the one hand, the explorations on the structural configuration, physical and chemical properties, and flow behavior of water constrained in nanoscale confinement are significant to some basic scientific researches, such as mass transfer under confinement and multi-field coupling at solid-liquid interface. On the other hand, the investigation on confined water demonstrates enormous potential applications in the fields of materials science, nanotechnology, water resource, environmental science, and energy utilization.

In this thesis, we performed MD simulations to investigate the phase transitions of 2D water constrained in graphene nanocapillaries. The lateral pressure, channel width, and temperature were three main factors considered in our simulations. We observed a myriad of 2D ice phases and associated phase transformations in MD simulations, and the compression-limit and superheating-limit (phase) diagrams were obtained. The main conclusions in this thesis are summarized as follows:

- Monolayer square-like ice (MSI), assembled square-rhombic units with grain boundaries, is a long-range ordered structure, in which the square and rhombic units are assembled in an order of alternative distribution, and the other rhombic unit forms grain boundaries. Spontaneous flipping of water molecules in MSI is intrinsic and induces transformations among different elementary units. Statistical averaging results demonstrate the inherent square characteristic in MSI.
- Four different phase transitions are found in the superheating process of monolayer ice in graphene nanocapillaries, including the two-stage melting of monolayer ice under ultrahigh lateral pressure. Monolayer triangular amorphous ice and two

Y. Zhu, *Phase Behavior of Two-Dimensional Water Confined in Graphene Nanocapillaries*, Springer Theses,
https://doi.org/10.1007/978-981-15-7957-8_8

different coexistence phases are obtained at ultrahigh lateral pressure and temperature. A schematic phase diagram of the melting of MSI is given to understand the transformations among different phases.

- AB-stacked bilayer square-like ice (BL-ABI) is a proton-ordered ice with square-like pattern. The phase transition from puckered MSI to BL-ABI at ambient temperature is through the crystal-to-amorphous-to-crystal pathway with a notable hysteresis-loop in the potential energy during the compression/decompression process, reflecting the compression/tensile limit of the 2D monolayer/bilayer ice.
- For AA-stacked bilayer ice phases, a temperature threshold (~275 K) is found between BL-VHDI and BL-AAI during the compression simulations of bilayer water. The compression and superheating of BL-VHDI display similar structural evolution; that is, square ice-nanotube arrays will bend first, followed by the transformation into BL-AAI. This structural transformation can be regarded as the buckling failure of square ice-nanotube columns, which is dominated by the high lateral pressure. In addition, in the transformations among different bilayer ices, solid-to-liquid-to-solid phase transition behavior (or Oswald staging) can arise due to that the hydrogen-bonding network must be disrupted first to reconstruct a new hydrogen-bonding network in the course of solid-solid transformation.
- Three (meta)stable trilayer crystalline/amorphous ice phases, namely, TL-ABAI, TL-ABA, and TL-AAAI, are observed in MD simulations. TL-ABAI can be regarded as a clathrate-like inerratic structure with a square lattice in each layer. For all three phases, the oxygen density profiles show that the peak of the middle layer is lower than those of the other two layers while the width of the middle layer is wider. The computed diffusivity suggests that water in the middle layer exhibits behavior different from that in the two outer layers in contact with the graphene. For TL-AAAI, water molecules in the layer next to graphene move faster than those in the middle layer.
- The compression limit of 2D water constrained in graphene nanocapillaries at ambient temperature is investigated systematically considering different channel widths. Lateral pressure can be used as the primary controlling variable to realize the first-order and continuous-like phase transitions in 2D confined water. The compression-limit (phase) diagram of 2D water constrained in graphene nanocapillaries at ambient temperature is obtained based on MD simulations.

8.2 Novelty

The novelty of this thesis is summarized as below:

- For the first time, we systematically investigated the compression limit of 2D confined water. When water is constrained between two sheets of graphene, it becomes a 2D liquid and then is turned into an intriguing monolayer solid with a square pattern due to the high van der Waals pressure. From a mechanics point of view, this liquid-to-solid transformation characterizes the compression limit (or metastability limit) of 2D liquid. We found that lateral pressure can be as a primary controlling parameter to realize the first-order and continuous-like phase transitions in 2D confined water. We revealed the influence of lateral pressure, channel width, and temperature on the structures and phase transitions of 2D confined in graphene nanocapillaries.

- We predicted a myriad of 2D crystalline/amorphous ice phases from MD simulations, including three monolayer phases (fMSI, pMSI and ML-T), seven bilayer phases (BL-ice I, BL-ABI, BL-AAI, BHI-AB, BL-AB_1 and BL-AB_2), and three trilayer phases (TL-ABAI, TL-ABA, and TL-AAAI). We demonstrated the inherent square characteristic of MSI obtained from MD simulations. ML-T, BL-ABI, BL-AAI, BHI-AB, TL-ABAI, TL-ABA, and TL-AAAI were first observed in our MD simulations, which enrich the knowledge of 2D water/ice.
- We found several pressure-dependent two-stage melting processes of 2D ices. The superheating of MSI is sensitive to the lateral pressure and channel width. Under ultrahigh lateral pressure, ML-T and coexistence phase would be arisen during the melting of MSI. The compression and superheating of BL-VHDI display similar structural evolution; that is, square ice-nanotube arrays will bend first, followed by the transformation into BL-AAI. For different bilayer ices, solid-to-liquid-to-solid phase transition behavior (or Oswald staging) can arise due to that the hydrogen-bonding network must be disrupted first to reconstruct a new hydrogen-bonding network in the course of solid-solid transformation.

8.3 Future Work

Although this thesis has carried out some investigations on the phase behavior of 2D water/ice, there are still some significant yet unresolved issues listed below:

- The existence of AA-stacked bilayer square ice at ambient temperature. The existence of AA-stacked bilayer square ice still remains a matter of debate. Although DFT calculations have obtained AA-stacked bilayer ice, this bilayer square structure is unstable and cannot be validated by MD simulations. There is a possibility that current theoretical simulations cannot capture the bilayer square ice structure or that the experimental measurements included other unknown factors. The discrepancy requires further investigation from both theoretical modeling and experimental measurements.
- The correlation between MSI melting and 2D-melting theory. 2D solids can melt via either a first-order phase transition or without any discontinuity in thermodynamic and may exhibit inter-mediate "x-atic" phases. The 2D-melting theory of two-step melting is based upon the behavior of topological defects in the form of dislocations and disclinations. Recent theoretical studies demonstrated that 2D melting for different systems exists complex phase transitions, including various phase coexistences. Monolayer ice is an special 2D material with abundance of hydrogen bonds. It is unsure whether monolayer ice could be discussed in the framework of 2D melting.

- The mechanisms of mass transfer under nanoscale confinement. Recent experiments demonstrated that graphene nanocapillaries in GO membrane exhibit excellent water permeability and intriguing ion-sponging effect. Graphene-based Angstrom-scale channel also exhibits highly effective water transport ability. Those abnormal mass transfer behavior under nanoscale confinement should be related to the phase behavior of 2D water and solid-liquid interfacial interaction at atomistic scale.

The manufacturer's authorised representative in the EU is Springer Nature Customer Service Centre GmbH, Europaplatz 3, 69115 Heidelberg, Germany. If you have any concerns regarding our products, please contact ProductSafety@springernature.com

Printed and bound by CPI Group (UK) Ltd, Croydon, CR0 4YY
30/09/2026
02225306-0005